THE PHYSICS OF CANCER

Recent years have witnessed an increasing number of theoretical and experimental contributions to cancer research from different fields of physics, from biomechanics and soft-condensed matter physics to the statistical mechanics of complex systems.

Reviewing these contributions and providing a sophisticated overview of the topic, this is the first book devoted to the emerging interdisciplinary field of cancer physics. Systematically integrating approaches from physics and biology, it includes topics such as cancer initiation and progression, metastasis, angiogenesis, cancer stem cells, tumor immunology, cancer cell mechanics and migration.

Biological hallmarks of cancer are presented in an intuitive and yet comprehensive way, providing graduate-level students and researchers in physics with a thorough introduction to this important subject. The impact of the physical mechanisms of cancer are explained through analytical and computational models, making this an essential reference for cancer biologists interested in cutting-edge quantitative tools and approaches coming from physics.

CATERINA A. M. LA PORTA is Associate Professor of General Pathology at the University of Milan and group leader of the Molecular Oncology Laboratory there. In 2015 she founded the university's Center for Complexity and Biosystems. She has organized several interdisciplinary workshops on the physics of cancer funded by CECAM and ESF.

STEFANO ZAPPERI is Professor of Theoretical Physics at the University of Milan, where he coordinates the Center for Complexity and Biosystems, and Research Leader at the ISI Foundation in Torino. He is an expert in the statistical mechanics of non-equilibrium complex systems and has worked in fracture, plasticity, friction, magnetism and biophysics. He is the recipient of numerous awards and is currently the chairman of the steering committee of the Conference on Complex Systems.

THE PHYSICS OF CANCER

CATERINA A. M. LA PORTA

Università degli Studi di Milano

STEFANO ZAPPERI

Università degli Studi di Milano

CAMBRIDGE
UNIVERSITY PRESS

CAMBRIDGE
UNIVERSITY PRESS

University Printing House, Cambridge CB2 8BS, United Kingdom

One Liberty Plaza, 20th Floor, New York, NY 10006, USA

477 Williamstown Road, Port Melbourne, VIC 3207, Australia

4843/24, 2nd Floor, Ansari Road, Daryaganj, Delhi – 110002, India

79 Anson Road, #06–04/06, Singapore 079906

Cambridge University Press is part of the University of Cambridge.

It furthers the University's mission by disseminating knowledge in the pursuit of education, learning and research at the highest international levels of excellence.

www.cambridge.org
Information on this title: www.cambridge.org/9781107109599
DOI: 10.1017/9781316271759

© Caterina A. M. La Porta and Stefano Zapperi 2017

First published 2017

Printed in the United Kingdom by TJ International Ltd. Padstow Cornwall

A catalogue record for this publication is available from the British Library.

Library of Congress Cataloging-in-Publication Data
Names: La Porta, Caterina, 1966– | Zapperi, Stefano.
Title: The physics of cancer / Caterina La Porta, Università degli Studi di Milano, Stefano Zapperi, IENI-CNR, Milan.
Description: Cambridge : Cambridge University Press, 2017. | Includes bibliographical references and index.
Identifiers: LCCN 2016057382 | ISBN 9781107109599
Subjects: LCSH: Eukaryotic cells. | DNA replication. | Cancer. | Cancer cells.
Classification: LCC QH447 .L37 2017 | DDC 571.6–dc23
LC record available at https://lccn.loc.gov/2016057382

ISBN 978-1-107-10959-9 Hardback

To Adriana and Ginevra with love.

Contents

Preface

Understanding how cancer initiates, grows and migrates has been a fundamental topic of biomedical research in the past decades and is still the object of intense scientific activity. Cancer is a complex disease where many factors cooperate. According to the 2000 seminal paper by Hanahan and Weinberg, six biological capabilities (the *hallmarks of cancer*) encapsulate the key features describing the remarkable variability displayed by cancer: sustaining proliferative signaling, evading growth suppression, activating invasion and metastasis, inducing angiogenesis, resisting cell death, activating invasion and metastasis, enabling replicative immortality (Hanahan and Weinberg, 2000). In a more recent review, the same authors take into account the observations that emerged in the previous ten years and add four new hallmarks: avoiding immune destruction, promoting inflammation, genome instability and mutation, deregulating cellular energetics (Hanahan and Weinberg, 2011). Hence, after ten years the hallmarks of cancer are still to be understood and have even increased in number! This is a signal in our opinion that traditional approaches need new strings in their bows. Tumors are extremely heterogeneous and their growth depends on dynamical interactions among cancer cells and between cells and the constantly changing microenvironment. All these interactive processes act together to control cell proliferation, apoptosis and migration. There is an increasing consensus that these dynamical interactions cannot be investigated purely through single biological experiments because experimental complexity usually restricts the accessible spatial and temporal scales of observations. Therefore it is necessary to study cancer as a complex system.

Advances in systems biology are already beginning to have an impact on medicine through the use of computer simulations for drug discovery. The integration of new experimental physics techniques in cancer research may help improve, for example, the capability to design more efficient cancer therapies. Understanding the biology of cancer cells and the impact of physical mechanisms, such as cell and tissue mechanics, on their biological functions could help validate biomarkers and develop more accurate diagnostic tools and individualized cancer therapies. In the past years, clinical studies have relied heavily on conventional population-based randomized clinical trials that try to identify favorable outcomes as an average over the population. Cancers are, however, extremely heterogeneous even within the same tumor class, so that an average positive outcome does not necessarily translate to a positive outcome in individual cases. A novel interdisciplinary quantitative approach to cancer where physics and biology work in an integrated manner could help in explaining the main steps involved in cancer progression and guide individualized therapeutic strategies (Deisboeck et al., 2011).

In past years, contributions to cancer research arose sporadically from different fields of physics, including, among others, biophysics, biomechanics, soft-condensed matter physics and the statistical mechanics of complex systems. Cell biophysics is becoming a mature field, both experimentally and theoretically, but contacts with forefront cancer research have been only intermittent. It is becoming clear that mechanical forces play a pivotal role in many fundamental biological functions such as cell division and cancer processes, like metastasis. Biophysical measurements that compare the mechanical properties of normal and cancer cells have consistently shown that cancer cells are softer than normal cells and that this compliance correlates with an increased metastatic potential. This is related to the fact that a softer cytoplasm corresponds to a less organized cytoplasm. Soft-condensed matter studies materials such as colloids and polymers that have a direct relevance for biology, and the distance between the two fields is rapidly shrinking. Finally, methods derived from the statistical mechanics of complex non-equilibrium systems have been applied to a wide variety of biological problems, ranging from protein folding and the analysis of genetic data, to the spreading of epidemics, but applications to cancer are still rare. These extremely promising and innovative research activities are currently shaping a new scientific field based on the physics of cancer.

Despite the fact that interesting novel contributions to cancer research grounded in physical sciences are emerging, they are so far mostly ignored by mainstream cancer research, which relies on the traditional tools of biologists, such as biochemistry and genetics. Over the years, biologists have turned to engineers or computational scientists to solve specific problems, but their role has always been somewhat ancillary. The complexity of this issue would require instead a truly interdisciplinary approach where physical scientists and engineers will become partners with biologists and clinicians in defining and addressing challenges in medicine. A similar approach could lead to fundamental advances in a wide variety of fields, such as medicine and health care, but it will only flourish when solid interdisciplinary education is finally available to students in science and medicine

(Sharp and Langer, 2011). The present book on the physics of cancer tries to contribute in filling this gap by providing a general introduction to an emerging and challenging interdisciplinary field.

The book is conceived as an introductory study concerning the physics of cancer. Its targeted readership is composed of graduate students in physics, biology and biomedical research, especially those focusing on quantitative and/or computational biology. The book should also appeal to established researchers in physical, biological and biomedical sciences who wish to enter this new field. We have thus engaged in the admittedly complex task of making the book accessible to both biologists and physicists. To this end, we strived to provide didactic explanations of basic concepts and ideas both in physics and cancer biology, avoiding excessive jargon and recourse to heavy mathematical formalism.

The book is structured into eleven chapters addressing key aspects of cancer research from an interdisciplinary perspective. Chapter 1 provides a general introduction to the biology of the cell that should be especially useful for physicists and engineers. We also discuss biophysical aspects of the cell that could be useful for biologists who may already be familiar with these topics but from a different perspective. Chapter 2 describes and discusses the biological mechanisms underlying cancer from its origin to its progression, focusing on the main processes involved such as angiogenesis and metastasis. Chapter 3 provides a discussion of modeling strategies employed to study cancer. The chapter first starts with a simple introduction to the theory of branching process and its application to cell proliferation and then continues with more involved mathematical and statistical physics models for cancer evolution and progression. We also discuss numerical simulations of metabolic pathways. While some background in mathematics and probability is needed to fully appreciate this chapter, we have worked to make the topic understandable to biologists. Chapter 4 examines in more depth angiogenesis, describing the concept of vasculogenic mimicry and its consequences for cancer development. Physics-based computational models of angiogenesis are critically discussed. A new frontier in cancer research is focusing on the presence of a more aggressive cell subpopulation, known as cancer stem cells; Chapter 5 discusses our current understanding of cancer stem cells, including stochastic models that describe their kinetics and evolution. We also point out methodological problems involved in their detection. Chapter 6 is devoted to the role of mechanical forces in tumor growth, both for individual cells and for a tumor mass inside tissues. We discuss experimental measurements of mechanical properties of cancer cells. Chapter 7 is devoted to cell migration and its role for metastasis. Chapter 8 discusses chromosomal instability and nuclear mechanics. Most tumors are immunogenic, and a possible strategy to fight the cancer is to use immunity against cancer. In Chapter 9, we provide a biological overview both of the role of immunity in cancer and a discussion of computational models to study this kind of inter-action. In Chapter 10, we discuss new challenges coming up in recent years on therapeutic strategies employed to fight cancer. We discuss classical methods based on radio- and/or chemo-therapy and more recent advances grounded in the physical sciences and nanotechnologies. The final chapter is devoted to a general outlook of the problem, focusing on how to integrate biological and physical sciences to build a new interdisciplinary field of research and train a new generation of scientists.

Acknowledgments

We would like to thank all the colleagues and collaborators who have contributed to shape our understanding of the physics of cancer. In particular, we wish to acknowledge interesting discussions with Mikko Alava, Martine Ben Amar, Eytan Domany, Ben Fabry, Lasse Laurson, Heiko Rieger, Jim Sethna, Alessandro Taloni, Guido Tiana and Lev Truskinovsky. Aalto University and the Weizmann Institute of Science are gratefully acknowledged for their hospitality, providing a quiet but stimulating environment where parts of this book were written.

1

Introduction to the Cell

In this chapter we will introduce the main properties of the eukaryotic cell, starting from its characterization in terms of its organelles, such as the nucleus, and its structural components, such as the cytoskeleton and the membrane (see Section 1.1). In Section 1.2, we discuss in detail how DNA is organized, and we introduce different chromatin structures such as B-DNA and Z-DNA and their condensation into chromosomes. Section 1.3 discusses how DNA is replicated so that the genetic information it encodes can be passed over to daughter cells. We also explain how DNA is repaired when damage due to external perturbation occurs. Next, in Section 1.4, we explain how the genetic information encoded in the DNA is transcribed into RNA and then translated into proteins, a process that has been termed the "central dogma" of molecular biology. Cells are surrounded by the plasma membrane formed by lipid bilayers, which also enclose the organelles and have a key role in intracellular and extracellular transport. This issue is illustrated in Section 1.5. Section 1.6 discusses the regulation of gene expression in the cell, and in particular that performed by miRNAs, a set of small RNA molecules. Finally, Section 1.7 illustrates the process of cell division and Section 1.8 discusses cell death and cell senescence.

1.1 Architecture of the Eukaryotic Cell

A cell is a small organized machine where DNA stores information, RNAs translate the message in protein language and proteins are the effectors. The ingredients needed to control the behavior of a cell are a mixture of biochemical and physical factors. Before discussing in detail how the machine functions, we first describe its general architecture. Cells in the human body can differ widely in terms of shape, size and function, but some general features are common to all cell types and are illustrated in Figure 1.1.

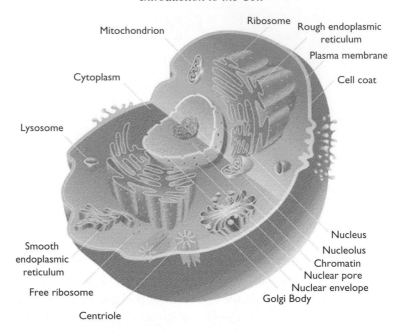

Figure 1.1 A schematic representation of an eukaryotic cell, displaying the organelles. Image by Mediran, CC-SA-BY-3.0 licence.

Eukaryotic cells are enclosed by the plasma membrane, a semi-permeable membrane in the form of a lipid bilayer that we will discuss in detail in Section 1.5. In contrast to bacterial cells, eukaryotic cells contain a set of membrane bound structures, known as organelles, that perform specialized functions. The largest cell organelle is the nucleus, which has a diameter that ranges between 3 to 10 μm and stores most of the cell DNA (Figure 1.2a). The nucleus is bound by a nuclear envelope composed of a protein network, enclosed in a bilayer membrane. Nuclear pores allow the passage of materials into and out of the nucleus into the cytoplasm which includes other organelles and the fluid cytosol.

The nucleus is surrounded by the endoplasmic reticulum (Figure 1.2c), a complicated maze of membrane-bound cisternae where ribosomes synthesize proteins. Proteins are shipped by vesicles to the Golgi apparatus, which looks like a set of layers of flattened disks and is the site of protein sorting (Figure 1.2d). From the Golgi apparatus, proteins are then transported to other regions of the cells. Mitochondria are shaped like cylinders with rounded ends, with a diameter of around 0.5 μm, and bound by a double membrane (Figure 1.2b). They are the site of production for adenosyne triphosphate (ATP), which provides the energy source for many cellular processes.

Cells display a complex mechanical structure provided by the cytoskeleton, a network of filaments of varying size and elastic properties (Figure 1.3); for

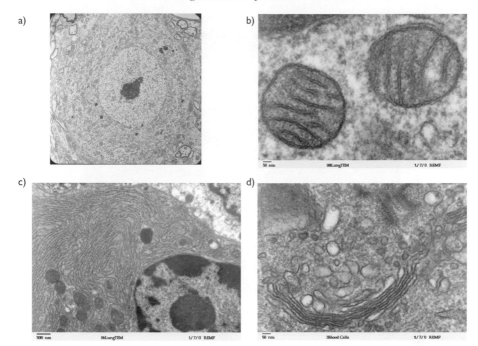

Figure 1.2 Cell organelles from transmission electron microscopy: a) A transmission electron micrograph of an eukaryotic cell (motor neuron from the bone marrow) showing the nucleus. Image by M. E. Pasini and M. Gioria. b) Mammalian lung cell showing mitochondria. c) Mammalian lung cell showing the nucleus and the rough endoplasmic reticulum. d) The Golgi apparatus from a red blood cell. Images b)–d) by Louisa Howard and Miguel Marin-Padilla, public domain.

instance, thin actin filaments, with diameter 7nm, often attached to the plasma membrane; a wide variety of intermediate filaments, with diameters around 10nm; and thicker and rigid microtubules, of diameter 25nm, radiating from the microtubule organizing center (MTOC). Microtubules play a key role in cell division.

1.2 The Organization of Genetic Material (DNA, Chromosomes, Genomes)

A DNA molecule is composed of two strands, held together by the hydrogen bonding between four bases: adenine (A), cytosine (C), guanine (G) and thymine (T). Adenine can form two hydrogen bonds with thymine; cytosine can form three hydrogen bonds with guanine. Although other base pairs (e.g., (G:T) and (C:T)) may also form hydrogen bonds, their strengths are not as great as those of (C:G) and (A:T) found in natural DNA molecules. Due to the specific base pairing, DNA's two strands are complementary to each other. Hence, the nucleotide sequence of one strand determines the sequence of the other strand. By convention, the sequence in

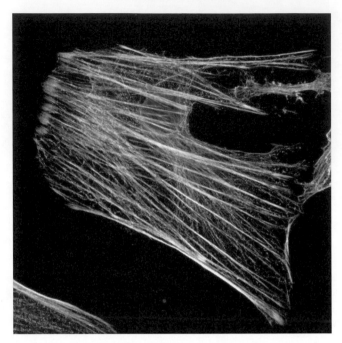

Figure 1.3 The cytoskeleton of HeLa cervical cancer cells. The fluorescent image
shows actin filaments, microtubules and the cell nucleus. Image by S. Wilkinson
and A. Marcus, National Cancer Institute, public domain.

a DNA database refers to the sequence of the 5' to 3' strand (left to right). Figure
1.4a shows the interaction between the nucleotides.

In a DNA molecule the two strands are not parallel, but intertwine with each
other. Each strand looks like a helix and the two strands form a "double helix"
structure, as discovered by James D. Watson and Francis Crick (Watson and Crick,
1953). In this right-handed structure, also known as the B form, the helix makes a
turn every 3.4 nm, and the distance between two neighboring base pairs is 0.34 nm.
Hence, there are about 10 pairs per turn. The intertwined strands form two grooves
of different widths, referred to as the major groove and the minor groove, which
may facilitate binding with specific proteins (see Figure 1.4b). In a solution with
higher salt concentration or with alcohol added, the DNA structure may change to
the A form, which is still right-handed, but making a turn every 2.3 nm with 11
base pairs per turn. Another DNA structure is called the Z form, because its bases
resemble a zigzag. Z-DNA is left-handed (see Figure 1.4b). One turn spans 4.6 nm,
comprising 12 base pairs. DNA molecules with alternating G-C sequences in alcohol or high salt solutions tend to have such a structure. The biological relevance of
Z-DNA has attracted much attention in recent years. In fact, the Z-conformation
was believed to trap the negative supercoiling when transcription occurs (Ha et al.,

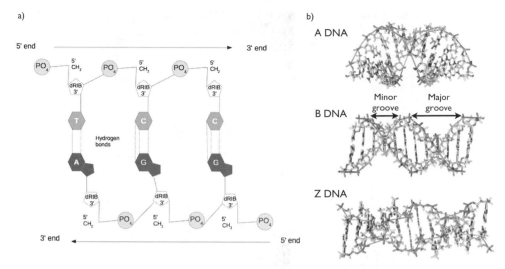

Figure 1.4 Structure of DNA. a) Interaction between nucleotides along DNA. b) DNA double helix in B and Z forms, differing by their chirality. The B form of DNA is characterized by an alternation between minor and major grooves. Adapted from an image by Richard Wheeler CC-BY-SA-3.0 licence.

2005; Rich and Zhang, 2003). Recent evidence shows a new role in gene transcription of the Z-DNA binding motif editing enzyme ADAR12529 (Kim et al., 1997). Moreover, Z-DNA is shown to be highly hydrophobic (Li et al., 2014).

The DNA duplex has two negatively charged phosphate backbone strands spiraling around the middle core of nucleotide base pairs (Figure 1.4b). Due to the high negative charge of this polyelectrolyte, the inter-DNA interaction is electrostatically repulsive. However, meters-long genomic DNA is packed by nature into compact structures in all living beings. To condense DNA, attractive forces must overcome repulsive forces. In most eukaryotic cell nuclei, DNA is packed in the form of chromatin by forming a complex with specialized histone proteins. Both analytical theories and computer simulations have been carried out to reveal the origin of like-charged DNA–DNA attraction and the roles of cations near DNA (Rouzina and Bloomfield, 1998; Shklovskii, 1999; Korolev et al., 2010). Generally, DNA–DNA interaction is caused by ion fluctuations and charge-bridging effects and is cation-dependent.

According to the most accepted model of DNA packing, chromosomes are hierarchically coiled: first, DNA is coiled around nucleosomes; second, the nucleosome strand is coiled into a 30 nm fibre; third, the 30 nm fibre is coiled into higher-order loops filling the volume of the chromosome (see Figure 1.5). A nucleosome consists of approximately 147 base pairs of DNA wrapped around eight histone protein cores. Linker DNA, more than 80 base pairs long, connects two histones between each nucleosome unit.

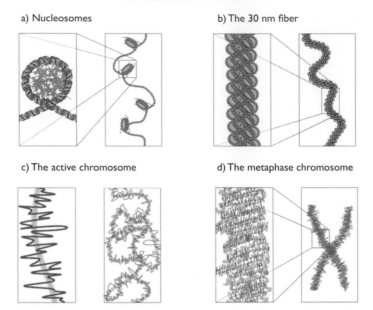

Figure 1.5 Structure of chromosomes. a) DNA is packed around histones, forming nucleosomes that are arranged like beads on a string. b) The nucleosomes are
arranged in a 30 nm fiber. c) During interphase, chromosomes in the chromatin
form long polymer-like structures thanks to scaffolding proteins. d) During cell
division, in metaphase, chromatin condenses into chromosomes. Adapted from an
image by Richard Wheeler, CC-BY-SA-3.0 licence.

The structure of the 30 nm fibre has not been resolved. There are two com-
peting models proposed on the basis of in vitro data: the solenoid model and the
zigzag arrangement of nucleosomes. In the one-start solenoid model, consecutive
nucleosomes interact with each other and follow a helical trajectory with bending
of linker DNA. In the two-start zigzag structure, two rows of nucleosomes form
a two-start helix so that alternate nucleosomes become interacting partners, with
relatively straight linker DNA. Twisting or coiling of the two stacks can produce
different forms of the zigzag model. Recently, the existence of the 30 nm fibre has
been questioned, due to the lack of reproducible observations in mammalian cells
(Nishino et al., 2012), leading to alternative models (Fussner et al., 2011).

Artificial segments of chromosomes were created and tagged by the insertion
of multiple copies of the lac operator sequence (Strukov and Belmont, 2009). The
idea was to distinguish whether the same nucleosomes are always positioned in the
same 3D space within a metaphase condensed chromosome. The results showed
no reproducibility in the lateral positions of the tagged sequences in mitotic chromosomes, even in sister chromatids. These experimental results do not agree with
a reproducible hierarchical folding model for the chromosome, suggesting instead
a disordered compaction of chromatin at metaphase, or at some other stage in the

condensation process. In summary, the structure of chromatin is heterogeneous both locally and globally.

A proteinaceous chromosome scaffold was believed to anchor DNA loops. However, in 2002, Poirier and Marko measured the physical properties of chromosomes by holding them between two micropipettes, showing that chromosomes are elastic and that DNA contributes to the elasticity as much as the proteins (Poirier and Marko, 2002). These results preclude the possibility that chromosomes are shaped by a proteinaceous scaffold, and suggest instead that they are formed by a DNA–protein meshwork. These data were then confirmed by electron microscopy tomography of imaged snap-frozen chromosomes (König et al., 2007). Therefore, chromatin fibres criss-cross the chromosome and are interlinked at frequent intervals. The condensin protein complex is found at these interlinks, suggesting that this fundamental chromosome constituent may function as a crosslinker of the meshwork. More recently, small angle X-ray scattering showed that nucleosomes do not display a folding hierarchy (Nishino et al., 2012). The current view is that the chromosome is the result of a self-organizing nucleosome chain constrained by condensin-mediated interactions.

1.3 DNA Replication, Repair and Recombination

The information enclosed in the DNA should be passed to the daughter cells avoiding missing or changed content. The machine that duplicates DNA is called replication. From the biochemical point of view, DNA replication is an intricate process requiring the concerted action of many different proteins. Because each resulting DNA double helix retains one strand of the original DNA, DNA replication is said to be semi-conservative. We will not enter here into the details of the biological factors involved, since there are many books available on the subject, but focus instead on the open problems that lie at the boundary between physics and biology.

To begin the process of DNA replication, the two double helix strands are unwound and separated from each other by the helicase enzyme. The point where the DNA is separated into two single strands is known as the replication fork. This is also the place where new DNA will be synthesized. After the single strands are exposed, they are rapidly coated by single-strand binding proteins (SSBs). SSBs keep the strands separated during DNA replication, preventing them from snapping back together. SSBs are weakly bound to the DNA and they are displaced when the polymerase enzymes begin synthesizing new DNA strands. Once separated, the two single DNA strands act as templates for the production of two new, complementary DNA strands. Remember that the double helix consists of two antiparallel DNA strands with complementary $5'$ to $3'$ strands running in opposite directions. Polymerase enzymes can synthesize nucleic acid strands only in the direction running from 5′ to 3′, hooking the 5′ phosphate group of an incoming nucleotide onto the 3′ hydroxyl group at the end of the growing nucleic acid chain.

When the strands are separated, DNA polymerase cannot simply begin copying the DNA. In fact, DNA polymerase can only extend a nucleic acid chain but cannot start one from scratch. To give DNA polymerase a place to start, an RNA polymerase called primase first copies a short stretch of the DNA strand. This creates a complementary RNA segment, up to 60 nucleotides long, called a primer. Now DNA polymerase can copy the DNA strand. The DNA polymerase starts at the 3′ end of the RNA primer, and, using the original DNA strand as a guide, begins to synthesize a new complementary DNA strand. Two polymerase enzymes are required, one for each parental DNA strand. Due to the antiparallel nature of the DNA strands, however, the polymerase enzymes on the two strands start to move in opposite directions. One polymerase can remain on its DNA template and copy the DNA in one continuous strand. However, the other polymerase can only copy a short stretch of DNA before it runs into the primer of the previously sequenced fragment. It is therefore forced to repeatedly release the DNA strand and slide further upstream to begin extension from another RNA primer. The sliding clamp helps hold this DNA polymerase onto the DNA as the DNA moves through the replication machinery. The continuously synthesized strand is known as the leading strand, while the strand that is synthesized in short pieces is known as the lagging strand. The short stretches of DNA that make up the lagging strand are known as Okazaki fragments.

The information contained inside DNA can be altered by mutations that could be induced by UV light, the most important source of DNA damage, or by the action of other chemical agents or errors during replication. These errors can be corrected by pre-replication or post-replication repair. Genetic information can be stored stably in DNA sequences only because a large set of DNA repair enzymes continuously scan the DNA and replace any damaged nucleotides. DNA repair mechanisms rely on the presence of a separate copy of the genetic information in each of the two strands of the DNA double helix. An accidental lesion in one strand can therefore be removed by a repair enzyme. Afterwards, a correct strand is resynthesized using the undamaged strand as a reference. Most of the DNA damage is removed thanks to one of two major DNA repair pathways. In base excision repair, the altered base is removed by a DNA glycosylase enzyme, followed by excision of the resulting sugar phosphate. In nucleotide excision repair, a small section of the DNA strand surrounding the damage is removed from the DNA double helix as an oligonucleotide. In both cases, the gap left in the DNA helix is filled in by the sequential action of DNA polymerase and DNA ligase, using the

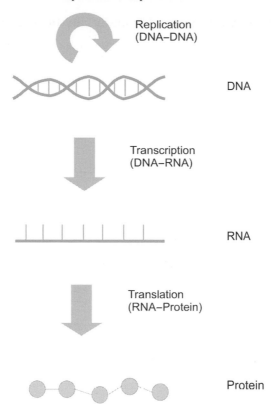

Figure 1.6 Illustration of the central dogma of molecular biology. DNA is both replicated into DNA and transcribed into RNA. RNA is translated into proteins.

undamaged DNA strand as the template. Other critical repair systems, based on either non-homologous or homologous end-joining mechanisms, reseal the accidental double-strand breaks occurring in the DNA helix. In most cells, an elevated level of DNA damage causes both an increased synthesis of repair enzymes and a delay in the cell cycle. Both factors help to ensure that DNA damage is repaired before a cell divides. In general the process is quite efficient and the rare errors result in specific non-common genetic pathologies such as the ataxiatelangiectasia (AT), whose symptoms include neurodegeneration, predisposition to cancer and genome instability.

Exchange of genetic information between two molecules of DNA leads to genetic recombination whose result is the production of a new combination of alleles. In eukaryotes, during meiosis, recombination contributes to exchange of DNA between people. Recombination can, however, also occur during mitosis involving the two sister chromosomes formed after chromosomal replication. Genetic recombination can also occur in bacteria and archaea.

1.4 Transcription and Translation Machineries

The information stored into the DNA should be translated into a message compre-
hensible for the cell, leading eventually to a specific phenotype. This role is played
by the transcription and translation machineries. Transcription is the process by
which the information in DNA is copied into messenger RNA (mRNA) for protein production. Finally, the translation machinery translates the mRNA language
into proteins. It is important to remark that the amino acid code is degenerate and
each triplet of nucleotides corresponds to more than a single amino acid. From this
observation it follows that changes in the DNA sequence do not necessarily alter
amino acid composition.

A detailed description of the transcription and translation machineries is easy to
find in any biological textbook. Here we would like just to sketch a few interesting
points. Transcription begins with a bundle of factors assembling at the promoter
sequence on the DNA, until the arrival of the RNA polymerase that initiates transcription. These assembled proteins require contact with activator proteins that bind
to specific sequences of DNA, known as enhancer regions. Once a contact is made,
the RNA polymerase transcribes the gene by moving along the DNA. According
to a well-accepted scenario, stable enhancer-promoter chromatin loops between
regulatory regions (Tolhuis et al., 2002) would establish physical contacts with a
promoter in static chromatin configurations, so that regulatory inputs are the same
in all cells. Thus transcriptional control would result from the action of binding
molecules (Tolhuis et al., 2002). Alternatively, enhancer-promoter contacts could
be considered as random events occurring in a fluctuating structural environment
(Fussner et al., 2011; Nora et al., 2012). This view implies that transcriptional regulation would be intrinsically heterogeneous, contributing to the observed cell to cell
variations (Amano et al., 2009; Krijger and de Laat, 2013). Following this idea, a
recent paper suggests that the contacts between potential regulatory elements occur
in the context of fluctuating structures rather than stable loops (Giorgetti et al.,
2014).

1.5 Membrane Structure and Intracellular Trafficking

The plasma membrane surrounding the cell is formed by a lipid bilayer, whose
structure is revealed by high-magnification electron micrographs such as the one
shown in Fig. 1.7a, showing two dense lines separated by a space (Cooper, 2000).
The lines are due to binding to the polar head groups of the phospholipids of the
metallic atoms used in the preparation of the sample for transmission electron
microscopy. The electron-dense lines are separated by the electron-poor internal
portion of the membrane, containing hydrophobic fatty acid chains.

a) b)

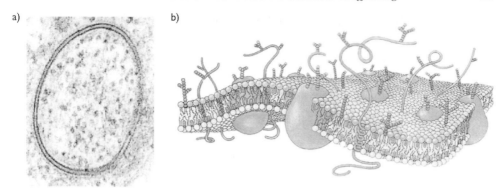

Figure 1.7 Structure of the plasma membrane. a) A transmission electron microscopy image of a lipid bilayer in a lipid vesicle. Public domain image. b) A schematic representation of the plasma membrane including the lipid bilayer and membrane inclusions. Image by William Crochot, CC-BY-SA-4.0 licence.

The plasma membranes of animal cells contain four major phospholipids: Phosphatidylcholine, phosphatidylethanolamine, phosphatidylserine and sphingomyelin. These molecules have an amphipathic character: they have hydrophilic heads and hydrophobic tails. Hence when phospholipids are placed in water they tend to form a bilayer structure in which the hydrophobic tails are screened from water which is only in contact with the hydrophilic heads. Phospholipids are differently distributed on the outer and inner leaflet of the plasma membrane of the cell. The inner leaflet contains mostly phosphatidylethanolamine and phosphatidylserine, but also a fifth phospholipid, phosphatidylinositol. The outer leaflet instead is mainly composed of phosphatidylcholine and sphingomyelin. Since the hydrophilic heads of both phosphatidylserine and phosphatidylinositol are negatively charged, the inner leaflet displays a net negative charge on the internal face of the plasma membrane. Notice also that phosphatidylinositol plays an important role in cell signaling despite the fact that its concentration inside the membrane is small.

In addition to the phospholipids, the plasma membranes of animal cells also contain a small amount (about 2% of the lipids) of glycolipids, found exclusively in the outer leaflet of the plasma membrane. On the other hand, an important membrane component of all animal cells is cholesterol which is present in about the same amounts as the phospholipids. Cholesterol is not present in bacteria and in plant cells, although they contain related compounds (sterols) that fulfil a similar function.

Phospholipid bilayers are of fundamental importance for membrane functions because they provide a barrier between two liquid compartments, being impermeable to most ions and biological molecules. Furthermore, lipid bilayers behave as soft elastic objects for out-of-plane deformations and as viscous liquids for in-plane shear deformations. This allows the transverse diffusion of phospholipids and proteins, a crucial property for many cellular functions. Not all proteins are, however, able to diffuse, since they are sometimes linked to the cytoskeleton, to proteins of neighboring cells or with the extracellular matrix (ECM).

Cholesterol is inserted into a bilayer of phospholipids with its polar hydroxyl group close to the phospholipid head groups. The role of cholesterol in bilayer membranes depends on temperature: at high temperatures it reduces membrane fluidity and permeability, while it has an opposite effect at low temperatures, maintaining membrane fluidity and avoiding freezing. An intriguing and still not completely understood phenomenon induced by cholesterol is the formation of "lipid rafts" in the fluid lipid membrane. Lipid rafts are cholesterol and sphingolipids enriched localized regions of the plasma membrane that are believed to play important roles in cell signalling and endocytosis, the process by which an extracellular molecule is surrounded by membrane and incorporated into the cell.

Membrane expansion and compartmentalization in eukaryotic cells lead to the development of a larger cell (from 1,000- to 10,000-fold increase in the volume) and an efficient specialization of the cellular function. It is of course obvious that good communication between the different compartments is fundamental. This goal is achieved using vesicles. Cargo-loaded vesicles form at a donor compartment with the help of specific coat and adaptor proteins (COPI, COPII and clathrin), then they are target to the appropriate receptor compartment to which they attach with the help of the SNARE protein complex (see Figure 1.8). Cells need to transfer material outward and inward; the first process is defined as the exocytic pathway and the second the endocytic one. In exocytosis, proteins synthesized in the endoplasmic reticulum (ER) are translocated to the plasma membrane. The rough ER is the site where all the secreted proteins are synthesized, as well as most lipids. From the ER, vesicles shuttle cargo towards the Golgi apparatus. In the trans-Golgi, proteins that should be secreted are packed into secretory vesicles and then fused with the plasma membrane in response to external stimuli. The Golgi cargo is sorted not only to the plasma membrane for regular secretion, but also to endosomes and lysosomes or back to the ER.

In the endocytic pathway, proteins and membrane are internalized early through a set of endosomes and late to the lysosomes, which are the major degradation sites of the cell. Cellular proteins can also go into the lysosomes through the autophagy and cytoplasmic to vacuole targeting pathway (CVT). There is also crosstalk between endocytic and exocytic pathways: endosomal and lysosomal resident proteins and enzymes are shuttled from the ER through the Golgi to endosomes and lysosomes and, in polarized cells, proteins can move from one side to

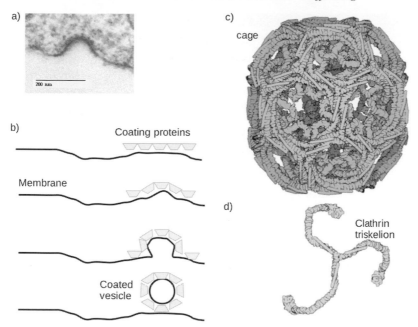

Figure 1.8 Coated vesicles. a) A coated membrane from transmission electron microscopy of a mouse embryo. Image by L. Howard and M. Marin-Padilla, public domain. b) A schematic representation of coated vesicle formation. First the membrane is coated by proteins, which induce curvature to the membrane until a coated vesicle is released. c) The structure of a clathrin cage is an assembly of d) triskelia, each composed of three clathrin units. Images c)–d) by David S. Goodsell and the RCSB PDB, CC-BY-4.0 license.

the other through the trans-cytotic pathway. Macromolecules can also be released from cells in small vesicles called exosomes by fusion of late endosomes (multivesicular bodies) with the plasma membrane. To maintain the compartmental size as well as the compartment identity, for each step of forward transport there is a retrograde transport where the membrane and the resident proteins are recycled back to the original compartments. The machinery that regulates this process is highly sophisticated.

One interesting and well-studied compartment is the Golgi apparatus, which displays a complex dynamics. How cargo moves through compartments, in particular the Golgi cisternae, is still an object of investigation. At present, the emerging idea is that cargo moves between compartments (ER, Golgi and plasma membrane) through vesicles; however, between sub-compartments vesicles are probably not the carrier of cargo. Further studies are needed to understand this aspect more deeply. Other intriguing unclear issues are the biogenesis of the intracellular compartment and the mechanism by which compartments are inherited into newly divided cells. During mitosis, the Golgi apparatus disassembles. In yeast, it seems

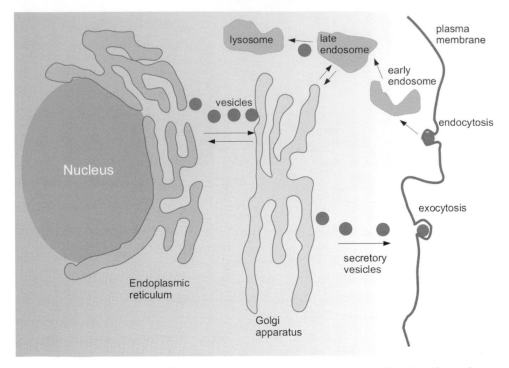

Figure 1.9 Vesicle trafficking. A schematic representation of endocytic and exocytic pathways, involving vesicle trafficking between the ER, the Golgi, endosomes and lysosomes.

that the Golgi is formed de novo without a template, while in mammals and protozoa the mechanism is self-assembly with a template (Lowe and Barr, 2007). All types of membrane fusion, with the only exception of mitochondria, require the formation of a protein complex, termed the SNARE protein complex, which is specific for each type of transport vesicle and target membrane (Söllner et al., 1993).

1.6 Cell Communication and miRNAs

Unicellular as well as multicellular organisms should communicate and interact with the environment, transmitting the information from the plasma membrane to the nucleus through biochemical signals, i.e. growth factors, which modulate the intricate network of signal transduction. Signal transduction pathways are interconnected networks, as shown in Figure 1.10. A huge literature describes in detail each pathway in specific cellular models, but the collective function of this network as a function of external stimuli is still an open question. In the era of Big Data, understanding how this network works is clearly a key issue requiring an interdisciplinary approach combining computer science and biology.

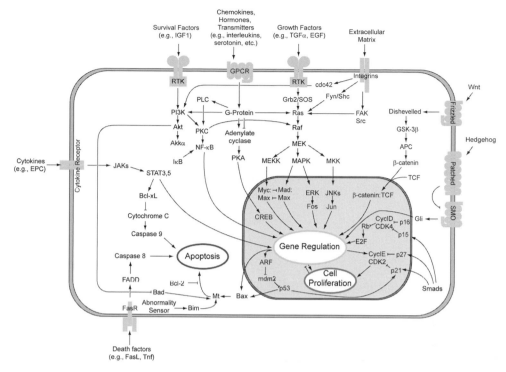

Figure 1.10 An illustration of the main signal transduction pathways in the human cell. Image by Roadnottaken, CC-BY-SA-3.0 licence.

Pathways can be activated by phosphorylation or dephosphorylation or by translocation of factors to intracellular compartments such as the nucleus. These changes can be due to the expression of transcription factors which regulate genes involved in key function of the cells. Due to the biological complexity of the signal transduction network, the output is not easy to predict. As shown in the KEGG metabolic pathway database, the signal transduction pathways are subdivided into functional subclasses: carbohydrate metabolism, energy metabolism, lipid metabolism, aminoacid metabolism, etc. Cells communicate with each other, releasing growth factors or cytokines which could affect the same cells if they express the receptor (autocrine stimulus) or nearby cells (paracrine stimulus). Some factors, such as hormones and some cytokines (i.e. IL1), can also act at long distance, reaching a target cell through the blood system. On the other hand, cells can communicate through the release of vesicles containing microRNAs (or miRNAs) (Valadi et al., 2007; Turchinovich et al., 2011), a large family of 22 nucleotide RNAs regulating many aspects of cell functioning (Martinez-Sanchez and Murphy, 2013).

miRNAs are mostly located in introns of both protein-coding and non-coding genes and are believed to be linked to the expression of other genes,

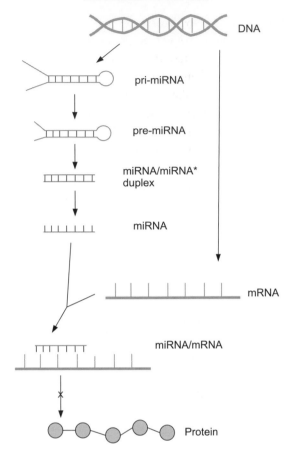

Figure 1.11 Regulation by miRNA. Silencing by miRNA is obtained by direct binding to mRNA target. miRNA is produced in stages from pri-miRNA, pre-miRNA and then miRNA duplex.

implying a coordination in the regulation of miRNA and protein expression. miRNAs are processed in two steps from precursor molecules (pri-miRNAs) transcribed by RNA polymerase II (see Figure 1.11). In the nucleus, the RNAse III enzyme DROSHA, in complex with other proteins, cleaves the pri-miRNA into a 70-nucleotide precursor (pre-miRNA) which is transported to the cytoplasm. In the cytoplasm, pre-miRNA is further processed by another enzyme acting in conjunction with an RNA-binding protein (TRBP in mammals), producing a small RNA duplex (20 nucleotides). One of the two strands is incorporated into a miRNA-induced silencing complex (miRISC) while the other one is released and degraded. In some cases, the second strand is not degraded but is also incorporated into miRISC, thus also functioning as a mature miRNA.

Regulation by miRNA occurs by its binding with complementary mRNA targets that are therefore silenced (Figure 1.11). In general, the most important region for target recognition is located in nucleotides 2–8 of the miRNA and binding sites are most commonly found in the 3′ untranslated region (UTR), the section of mRNA that immediately follows the translation termination codon, of the corresponding mRNAs. Many examples have been reported, however, where miRNA binding sites are located outside the 3′ UTR and are instead inside the coding region. Furthermore, miRNA can bind mRNA even when a perfect complementary is lacking, making it extremely hard to predict if a specific mRNA is regulated by a given miRNA. The task is typically performed by numerical algorithms trying to predict miRNA targets by seed pairing and evolutionary conservation. This results in the prediction of often hundreds, and sometimes thousands, of targets for each miRNA, with a large number of false-positives and false-negatives (Alexiou et al., 2009).

The fact that animal miRNAs are not perfectly complementary to their targets is the source of enormous computational challenges, since small differences in the algorithms can result in a large difference in target predictions. Hence, the typical strategy involved the use of multiple algorithms and subsequent validation by experimental methods. In general, algorithms relying on evolutionary conservation or the combination of multiple algorithms usually display high precision but low sensitivity (Alexiou et al., 2009). Furthermore, most current algorithms restrict their analysis to the 3′ UTRs, while, as noted above, in many cases miRNA/mRNA interactions occur through the coding sequence. In addition, miRNA/mRNA binding could be different in the 3′ UTR or other regions, further complicating the computational task.

An additional complication for miRNA target determination comes from the fact that miRNA/mRNA interactions are often tissue and context specific. A recent paper exploits this observation by combining traditional sequence-based prediction algorithms with paired miRNA and mRNA expression data (Bossel Ben-Moshe et al., 2012). By matching predicted targets with cross-correlations in miRNA/mRNA expression it is possible to considerably decrease detection errors. This further confirms that experimental validation of miRNA targets is essential and can be used to uncover additional rules of miRNA target binding, providing useful feedback for computational approaches. Since the functional role of miRNA is to inhibit their targets' mRNA, the most straightforward experimental approach for miRNA target detection is based on the transfection of specific miRNA inhibitors into the cells followed by high-throughput analysis of mRNA expression.

As we discussed in Section 1.5, cells may release membrane vesicles into their extracellular environment either by pinching them off directly from the plasma membrane or through secretion by endocytic compartments. Extracellular vesicles have been assigned several names, including microvesicles and exosomes. The term exosomes is generally coined for those vesicles that are secreted as a consequence of the fusion of multivesicular bodies with the plasma membrane. Multivesicular bodies are generated at endosomes by the inward budding of their delimiting membrane followed by the release of 100 nm vesicles into the endosomal lumen. Such intra-endosomal vesicles have a cytosolic-side inward orientation, just like cells. Many multivesicular bodies serve as a sorting station for endocytosed membrane proteins that need to be transferred to and degraded in lysosomes. Other multivesicular bodies may instead fuse with the plasma membrane, resulting in secretion of their intraluminal vesicles as exosomes. Exosomes are secreted by many if not most cell types, and are abundantly present in body fluids such as blood, ascites, urine, milk, saliva, and seminal plasma. Exosomes are proposed to have many distinct physiologic functions that may vary, depending on their cellular origin, from immune regulation to blood coagulation, cell migration, cell differentiation and other aspects of cell-to-cell communication. In 2007, Valadi and colleagues demonstrated that exosomes from mast cells contain both mRNA and miRNA, and that at least some of these mRNAs could be translated into proteins on their transfer by exosomes to target cells (Valadi et al., 2007). Since then, exosomes from many other cell types have also been demonstrated to carry RNA, as summarized in the database Exocarta.org. Multivesicular bodies are functionally linked to miRNA effector complexes, possibly indicating mechanisms for miRNA targeting to exosomes. The concept of exosome-mediated directed transfer of selected miRNA between cells is extremely attractive, although several basic elements of such a process still require confirmation. The idea is that a cell, in addition to the classical pathway of communication (i.e. growth factors), can communicate and affect other cells through miRNAs. In a recent paper, Montecalvo and colleagues demonstrate that dendritic cells (DCs) secrete exosomes that are loaded with distinct sets of miRNAs, dependent on the status of DC activation. Moreover, they show that DC exosomes can fuse with target cells, thereby delivering their membranous and cytosolic contents (Montecalvo et al., 2012). This is the first evidence of the intriguing possibility that cells use exosomes to deliver information.

1.7 Cell Division

The life of an eukaryotic cell follows a cycle punctuated by mitosis, a relatively short phase where the cell divides into two new cells, and a long interphase where the cell grows. The cell cycle is schematically illustrated in Figure 1.12: After mitosis, a cell starts to grow in size during the so-called first gap phase or G1. After the G1 phase, the cell undergoes a checkpoint for DNA damage. If the checkpoint is passed, the cell enters into the S phase where the process

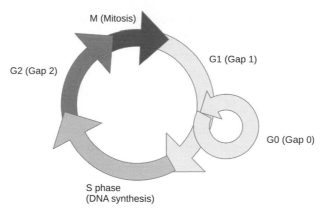

Figure 1.12 An illustration of the eukayotic cell cycle. Cells undergo a life cycle ending with their division (mitosis). A newly divided cell first stays in a first gap phase (G1) where cell size grows, then enters into the S phase where DNA is duplicated. Then a second gap phase (G2) leads to increased cell size, terminating again in mitosis. Some cells turn quiescent and enter the G0 phase where they do not grow.

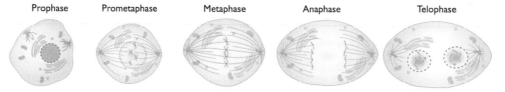

Figure 1.13 An illustration of the different phases of mitosis. Public domain image.

of DNA replication takes place as discussed in section 1.3. After this follows a second and shorter gap phase, or G2, where the cell grows again in size until it is ready for cell division. Differentiated cells sometimes exit from the cell cycle and enter into the G0 phase, where they do not grow, and remain quiescent.

Cell division is a fascinating, complex, tightly regulated process in which chromosomes must be correctly segregated into the two daughter cells thanks to the coordinated actions of a number of microtubules and protein motors. Mitosis is conventionally subdivided into different phases as illustrated in Figure 1.13. In the *prophase*, chromatin condenses into chromosomes. Genetic materials has already replicated during the S phase and therefore each chromosome is composed by two sister chromatids attached through the centromere, a protein complex that acts as the assembly point for the kinetochore. The kinetochore will play a central role in the following since it is the place where microtubules attach. During prophase, the centrosomes, the microtubules organizing centers, duplicate and start to form spindle poles. In *prometaphase*, the nuclear envelope breaks down and chromosomes become accessible to microtubules that are organized into two spindle poles. During prometaphase, chromosomes must align on a central plate between the spindle poles, a process known as congression (Magidson et al., 2011) and the kinetochores of all chromosomes should be attached to microtubules emanating from each of the two poles, thus becoming bi-oriented (Walczak et al., 2010). This process is guided by the action of microtubules that, thanks to their successive growth and shrinkage, search and capture chromosomes. Attached chromosomes are moved on microtubule tracks by molecular motors such as kinetochore dynein (Rieder and Alexander, 1990; Li et al., 2007; Yang et al., 2007; Vorozhko et al., 2008) and centromere protein E (CENP-E or kinesin-7) (Kapoor et al., 2006; Cai et al., 2009) and polar ejection forces (PEF) (Rieder et al., 1986), mainly originating from kinesin-10 (Kid) and kinesin-4 (Kif4A) motors that sit on the chromosome arms (Wandke et al., 2012).

The congression process ends in *metaphase* when chromosomes are all correctly aligned on the metaphase plate and attached through their kinetochores to microtubules from the two poles. In this condition, the kinetochore is under tension, otherwise the attachments are faulty and become de-stabilized by the action of the Aurora B kinase that resides in the centromere (Cimini et al., 2006; Lampson and Cheeseman, 2011). Only if this crucial spindle assembly checkpoint is passed will each chromosome divide into two chromatid sisters that are pulled synchronously towards the spindle poles during the *anaphase* (Matos et al., 2009). Finally, the cell enters into the *telophase* where proper cell division occurs, nuclei and other organelles are re-formed, and the cytoplasm is distributed among the two cells (cytokinesis).

1.8 Cell Death and Senescence

Every cell has a specific lifespan, a fact that is also useful to eliminate cells that are becoming non-functional, due, for example, to stress. Cell death mechanisms have traditionally been divided into two types: programmed cell death, or apoptosis, a mechanism that requires energy, and necrotic cell death, that does not. An important difference between these two types of death is that necrosis typically causes a strong immune response, whereas apoptosis does not.

A cell undergoing apoptosis displays an important shrinkage and condensation of nucleus and chromatin, called pyknosis, followed by their fragmentation, or karyorrhexis. Finally, one observes cell membrane blebbing (see Figure 1.14) and the budding of the cell into a series of membrane-bound structures, called apoptotic

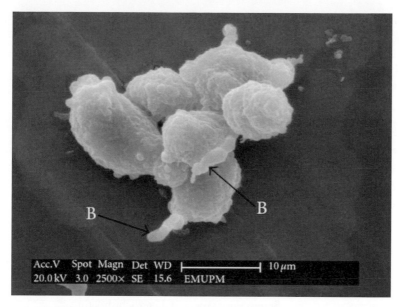

Figure 1.14 SEM micrographs of surface ultrastructural characteristics of HeLa cells undergoing apoptosis displaying blebs (B). Image by Abdel Wahab et al., 2009, CC-BY-3.0 licence.

bodies. The small size of apoptotic bodies allows their engulfment by nearby macrophages and other cells through phagocytosis, so that the material is broken down in phagolysosomes for recycling. Apoptosis does not cause any immune reaction since no cellular material and inflammatory cytokines are released into the interstitial space. The intracellular pathways involved in apoptosis are well known and studied (Tower, 2015).

The necrosis pathway for cell death is, in contrast, characterized by cell swelling and the disruptions of the cell membrane. Hence, the release of cytoplasm into the interstitial space typically results in inflammation. For this reason, necrosis is traditionally considered as a toxic process where the cell is passively destroyed, while apoptosis is believed to be an important process for optimal lifespan in species, such as mammals, where multiple tissues are maintained by a constant cell turnover.

Senescence represents another pathway used by cells before or instead of undergoing apoptosis (Rodier and Campisi, 2011; Campisi, 2013; López-Otín et al., 2013). Senescent cells, in contrast with quiescent ones, do not proliferate even when growth factors are added to the medium, as originally shown by Hayflick and Moorhead (1961). In aging organisms, excessive stress accumulation induces senescence in stem and progenitor somatic cells as well as mitotically active cells within renewable tissues. The senescence program is initiated by cell cycle arrest and results in the decline of proliferation and mobilization to renewable and

Table 1.1 *Features commonly associated with senescent cells (Piano and Titorenko, 2015)*

Affected aspect of cell morphology and function	Feature of senescent cell
Cell size and shape	Cell enlargement and acquisition of a flat or spindle-like shape
Cell Cycle	Cycle arrest in vivo; in vitro can be reversible under genetic manipulation
β-galactosidase	High activity
Mitochondria	Excessive proliferation of mitochondria changed morphology, depolarization, reduced ATP and accumulation of ROS
DNA damage foci	Permanent establishment of nuclear foci marked with a set of the DNA damage response
Nucleus bodies	Formation of promyelocytic leukemia nuclear bodies
Heterochromatic DNA foci	Senescent associated heterochromatin

differentiated tissues, which could compromise tissue repair and regeneration, and therefore impair tissue homeostasis (Rodier and Campisi, 2011; Campisi, 2013; López-Otín et al., 2013). Senescent cells in culture or in vivo undergo various morphological and functional changes, acquiring a number of features as illustrated in Table 1.1 (Piano and Titorenko, 2015). Some of these features are often observed in different types of cultured cells exposed to various senescence triggers and in senescent cells derived from tissues. These features could be used as hallmarks of a state of cellular senescence and as diagnostic biomarkers of cells entering such a state in different tissues. The most reproducible biomarker currently used for senescence is the β-galactosidase activity.

There is no consensus on the specific molecular and biochemical causes for aging and its relation to cell senescence. Aging is currently classified according to programmed versus damage or error-induced mechanisms. According to programmed mechanisms, aging is driven by a developmentally controlled program regulating tissue homeostasis and repair responses. Damage- or error-induced aging mechanisms emphasize instead the accumulation of stress-induced damage caused by reactive oxygen species, cross-linked macromolecules' DNA damage, or altered energy metabolism. Aging has also been explained by some as a direct consequence of the intrinsic thermodynamic instability of most complex biological molecules, leading to progressive loss of molecular fidelity (Hayflick, 2007).

2

The Biology of Cancer

Cancer results from abnormal cellular growth. It is considered to be benign when localised in situ while it is defined to be malign and metastatic when it is invasive and spreads inside the body through blood or lymphatic vessels. Cancer progression can be interpreted as an evolutionary process, as we discuss in Section 2.1. In spite of the increasing knowledge gained on the molecular mechanisms involved in the deregulation of cancer cells, such as the identification of many oncogenes and oncosuppressors (discussed in Section 2.2), many open questions still exist about the origin of cancer cells. In Section 2.3, we introduce a key oncosuppressor gene which is of fundamental importance for cancer development: P53, also known as the "guardian of the genome." While important oncogenes and oncosuppressors clearly exist, cancer involves a multitude of different genes requiring an integrative data-based approach (Section 2.4).

Another important issue that is still under investigation is the presence of a subpopulation of more aggressive cancer cells, usually described as cancer stem cells (Section 2.5) (CSCs). The molecular aspects related to the capability of cancer cells to receive nutrients from the environment through existing vessels, and the ability of the same cancer cells to induce vessel formation (angiogenesis), are two critical aspects of the biology of cancer that we illustrate in Section 2.6. Furthermore, in Section 2.7, we illustrate the spread of cancer cells inside the body in the metastatic process. All together these aspects will be discussed here, combining biological and physical viewpoints. In this perspective, the cancer ecosystem is the combination of physical forces and biochemical ingredients. Finally, the new diagnostic tools for the identification of a cancer cell are also discussed and critically reviewed (Section 2.8).

2.1 Cancer Origin and Evolution

It is now commonly accepted that cancer is the consequence of random mutations in cells. This general idea dates back to the pioneering observations of chromosomal abnormalities in cancer made by Boveri at the beginning of the twentieth century (Boveri, 1903). The concept gained further traction with the discovery, made by Muller in the twenties, that ionizing radiation is mutagenic (Muller, 1930). In the forties, Berenblum and Shubik discovered that chemical carcinogenesis was described by two stages: initiation by carcinogens and promotion by other chemicals (Berenblum and Shubik, 1949). It was again Muller in the fifties who proposed that cancer initiates as a result of multiple mutations in a single cell (Muller, 1964).

When it became clear that more than one mutation was needed for cancer, a natural question was to determine the precise number. This issue was first tackled in the fifties using epidemiological data (Armitage and Doll, 1954; Nordling, 1953). It was noticed that the death rate due to various types of cancer was scaling as a power of age, with an exponent ranging between 5 and 6, depending on the tumor (two examples are reported in Figure 2.1).

A simple statistical argument can be used to understand this behavior. If we assume that mutations in a cell occur randomly with rate R, the probability that the Nth mutation occurs at time t is simply given by (Armitage and Doll, 1954)

$$p_N(t)dt = \frac{R^N t^{N-1}}{(N-1)!}dt, \tag{2.1}$$

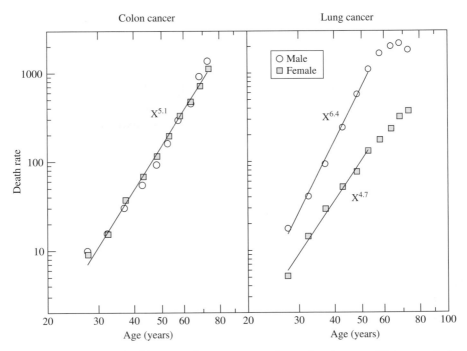

Figure 2.1 Cancer incidence scales as a power law related to age. Data reported for colon and lung cancer are obtained from Armitage and Doll (1954).

where the $(N-1)!$ normalization is introduced to avoid counting all the equivalent cases differing only by the order in which mutations occur. Hence, if we assume that cancer occurs after exactly N mutations, we would predict that its incidence should be proportional to t^{N-1}. This simple argument is derived under a number of strong assumptions like the fact that intermediate mutations do not confer any growth advantage to the cell, so that the cell population remains constant, or that the mutation rates do not depend on time. One can define more sophisticated statistical models to describe the evolution of mutations in a cell population (see, for instance, Bozic et al., 2010).

In Equation 2.1 it is assumed that the order in which mutation occurs is not relevant, but a large body of literature has been devoted to determining the exact order of mutations leading to cancer. For instance, in an influential paper, Fearon and Vogelstein (1990) reconstructed the order of mutations in colorectal cancer based on the analysis of cross-sectional data. This type of analysis suggests a linear evolution model for cancer where mutations accumulate one after the other. The reality is likely to be more complex, as shown by Sprouffske et al. (2011), who compared clinical data with the results of numerical simulations of individual cell-based models for cancer evolution.

The simple statistical arguments discussed above had a great influence in establishing the general idea that a small fixed number of mutations is needed to initiate cancer. Cancer could then be seen as the result of *clonal evolution*, in which cells would undergo a series of mutation and selection events, similar to the evolution of species (Nowell, 1976; Merlo et al., 2006; Greaves and Maley, 2012). As first discussed by Nowell, 1976, cancer would be initiated by mutations in one cell, which would then acquire some selective growth advantage and would then expand its lineage. As cells divided, they would acquire further mutations, some of them "deleterious" and leading to extinction, other "beneficial" and leading to more malignant phenotypes (driver mutations), and most of them just neutral (passenger mutations). The whole process could then be described as a tree, analogous to Darwin's tree of life (see Figure 2.2). This hypothesis is supported by histopathological evidence where cancer progression is identified by a series of well-defined stages (e.g., adenoma, carcinoma, metastases).

The general idea of cancer progression as an evolutionary process gained further confirmation with recent advances in molecular biology and genetics. Single nucleus sequencing now allows us to investigate mutations in a cell population coming from the same tumor (Navin et al., 2011; Anderson et al., 2011). For instance, Navin et al. (2011) analyzed 100 single cells from the same breast cancer and concluded that a single clonal expansion formed the primary tumor and the metastasis. Their analysis also supported the idea that cancer would evolve in

a)   b)

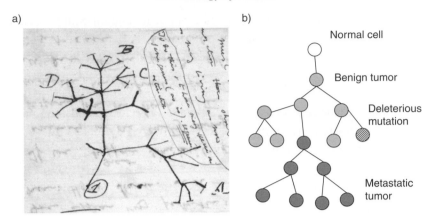

Figure 2.2 a) Darwin's tree of life (from First Notebook on Transmutation of Species, 1837). The tree represents the evolution of species. b) The cancer tree as first proposed by Nowell (1976): mutation in one normal cell leads to a benign tumor cell that replicates further until it acquires additional mutations, eventually becoming metastatic. Some mutations can be deleterious, leading to the extinction of a branch of the tree.

discrete bursts, as in the evolutionary theory of punctuated equilibria (Gould and Eldredge, 1993).

   We conclude this section by noting that, despite nearly a century of studies, the role of randomness in cancer initiation is still the subject of heated debates in the literature. A recent paper compared the incidence of each type of cancer to the number of stem cell divisions expected in the corresponding tissue (Tomasetti and Vogelstein, 2015). The authors found that the two variables were correlated and therefore concluded that most cancers are due to "bad luck," in that they would arise from random mutations in healthy tissues. The paper spurred great controversy since it seemed to downplay the well-established role of environmental factors and lifestyle for cancer incidence. A subsequent paper, for instance, showed that while the correlation exists, it cannot properly distinguish between intrinsic and extrinsic factors (Wu et al., 2015).

2.2 Oncogenes and Oncosuppressors

It is well established that cancer progression is induced by a small number of mutations involving two types of genes: oncogenes and oncosuppressors. Proto-oncogenes are genes that normally help cells grow. When a proto-oncogene mutates or it is present in too many copies, it becomes a "bad" gene and is called an oncogene. Oncogenes can be activated by translocation, point mutation, deletion, insertional activation or amplification. Since oncogenes are linked to critical biological functions of the cell, when they become permanently activated they induce an increase in cell proliferation.

A classical example is the activation of the gene Ras. All Ras protein family members belong to a class of protein called small GTPase, and are involved in signal transduction within the cell. The three Ras genes in humans (HRAS, KRAS, and NRAS) are the most common oncogenes in human cancers; mutations that permanently activate Ras are found in 20% to 25% of all human tumors and in up to 90% of certain types of cancer such as pancreatic cancer (Downward, 2003). The most important and well-known oncogenes belong to the following categories: growth factors, receptor tyrosine kinases, cytoplasmic tyrosine kinases, cytoplasmic serine/threonine kinases and their regulators, regulatory GTPase and transcription factors.

Oncosuppressors or tumor suppressor genes are genes that protect the cell, avoiding one crucial step on the path to cancer. When one of those genes mutates in a way that causes a loss or reduction in its function, the cell can progress to cancer, usually in combination with other genetic changes. Tumor-suppressor genes typically code for proteins that promote apoptosis, repress cell cycle regulation or are involved in DNA damage repair.

Unlike oncogenes, mutations in tumor suppressor genes usually lead to cancer only when both alleles coding for the proteins are affected (two-hit hypothesis). This is because if a single allele of the gene is mutated, the second can still correctly produce the protein. Oncogene mutations, in contrast, are usually effective even when they affect a single allele, because they just enhance the activity of the given protein. The two-hit hypothesis was first proposed by A.G. Knudson based on the study of 48 cases of retinoblastoma (Knudson, 1971, 2001). Knudson used Poisson statistics to show that the data are compatible with the assumption that retinoblastoma would occur after two independent random mutations (Knudson, 1971). In the hereditary version of the disease, one mutation is inherited from germinal cells and the other occurs in somatic cells. In the non-hereditary disease, both mutations occur in somatic cells. Using this general idea, Knudson fits the time dependence of the fraction of cases not diagnosed at time t. The curve scales as $\exp(-at)$ if a single random event is needed to initiate the tumor, while it would scale as $\exp(-bt^2)$ if two mutations were needed, as we will discuss in more detail in the next chapter. Here we report the original data with the theoretical fits (see Figure 2.3).

In a recent review, Hanahan and Weinberg proposed a revised version of their older *hallmarks of cancer* (Hanahan and Weinberg, 2000) where the capability of the cell to sustain proliferative signaling is discussed in the light of more recent biological discoveries. In particular, they revised the notion that an increasing expression of oncogenes would necessarily result in an increase of tumor

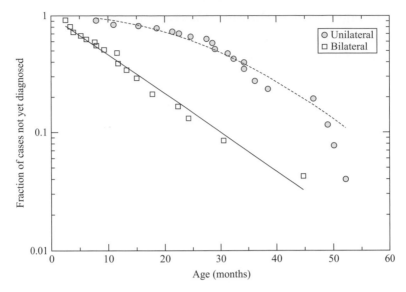

Figure 2.3 The fraction of cases of retinoblastoma not diagnosed at age *t* for bilateral and unilateral cases. The bilateral cases are supposed to be described by the hereditary version of the disease which should be triggered by a single mutation, while the unilateral cases are assumed to be non-hereditary and would then occur after two mutations. The fits are done according to the one mutation and two mutations models, respectively. Data are obtained from Knudson (1971).

growth (Hanahan and Weinberg, 2011). Recent data shows that cells can respond to strong signaling activity by oncoproteins, such as RAS, MYC and RAF, by inducing cell senescence and/or apoptosis (Collado and Serrano, 2010). Furthermore, experiments in mice engineered to lack telomerase show that shortened telomeres drive cells to senescence, reducing tumor formation even in mice genetically programmed to develop certain cancers (Artandi and DePinho, 2010). These apparently paradoxical responses could reflect an intrinsic cellular defense mechanism that would eliminate cells experiencing excessive signalling of certain types (Hanahan and Weinberg, 2011). This paradox is, however, still not resolved.

Another paradox is linked to the role of autophagy in tumors. Autophagy is an important physiological response of the cell that in normal conditions acts at low levels, similar to what happens with apoptosis. Under cellular stress, however, it can be strongly activated. There are conditions (like nutrient starvation or radiotherapy), however, in which elevated levels of autophagy become protective for cancer cells (White and DiPaola, 2009). Therefore in spite of the previous idea that everything regarding the origin of cancer cells was clear and easily explained by the activation oncogenes and the deactivation of oncosuppressors, the recent literature opens a more complex perspective. We can expect that our understanding of this issue may change completely in the near future.

2.3 The Role of P53 in Tumors

A key oncosupressor gene is p53, which has been described as the "guardian of the genome" (Lane, 1992). P53 belongs to a family which comprises p73 and p63 and is evolutionarily conserved in all animals. Functional and phylogenetic analyses reveal that the founding member was p63, followed by p73 and finally p53 (Yang et al., 1998, 2002). The importance of p53 stems directly from its central position in key cellular metabolic pathways (Figure 2.4), where it acts as a key regulator of the network (Vogelstein et al., 2000; Oren, 2003). The function of p53 is to respond to stress signals and DNA damage by inducing DNA repair or, in extreme cases, by sending cells into senescence or apoptosis. It turns out that in most cancers p53 does not function correctly, either because the gene itself is mutated or because MDM2, which negatively regulates p53 by promoting its ubiquitination and degradation, is overexpressed (Vogelstein et al., 2000). This implies that a central checkpoint in cells is evaded so that cancer cells can proliferate undisturbed.

The functioning of the p53–MDM2 pair is intriguing, since each of the two genes acts as as a regulator of the other: p53 induces MDM2, which in its turn inhibits p53 (see Figure 2.5, Lev Bar-Or et al., 2000). This kind of feedback loop is quite common in biological processes, leading to oscillations in gene expression and metabolite abundance, as shown by simulating nonlinear reactions in networks

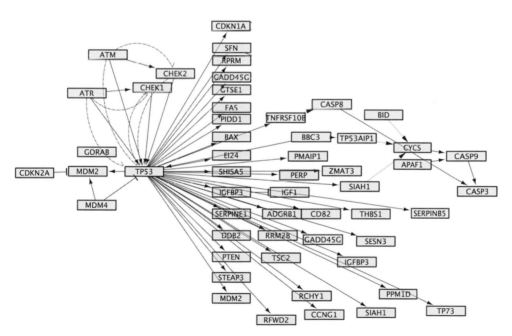

Figure 2.4 The p53 signaling pathway illustrates the centrality of p53 (here TP53) in the network. Image obtained with Cytoscape using KEGG pathways.

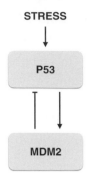

Figure 2.5 The feedback loop between p53 and MDM2. Stress activates p53 that induces MDM2, which inhibits p53.

containing positive and negative feedback loops and time delays (Tiana et al., 2007). Experimental studies of the dynamics of p53 and MDM2 proteins indeed confirm that their expression levels oscillate as a function of time (Lev Bar-Or et al., 2000; Geva-Zatorsky et al., 2006). Several mathematical and computational models have been proposed to explain the kinetics of the p53–MDM2 pair. The simplest models predict the presence of damped oscillations (Lev Bar-Or et al., 2000), but while damped oscillations are observed in cell populations (Lev Bar-Or et al., 2000), experiments in individual cells clearly show that the oscillations are instead undamped (Lahav et al., 2004). This theoretical issue can be explained (see Section 3.3.2) by considering transcriptional delays in the production of MDM2 when it is induced by p53 (Tiana et al., 2002) or by involving additional interme-diate components and feedback loops (Ciliberto et al., 2005). The most important question still to be clarified, however, concerns the functional role played by p53 oscillations and its effect on regulatory processes in the cell.

2.4 Big Data in Cancer

The last decades have seen the development of high-throughput or next-generation sequencing technologies, allowing researchers to obtain large amounts of data at relatively low cost for DNA sequences, transcriptome profiling, miRNA expres-sions, DNA–protein interactions and epigenetic classifications. This type of infor-mation is believed to be of particular relevance for cancer research, and for this reason international consortia are currently accumulating vast amounts of data on cancer profiling in public databases.

The Cancer Genome Atlas (TGCA `https://tcga-data.nci.nih.gov/tcga/`) maintained by the US National Cancer Institute collects genomic data for dozens of tumor types obtained from thousands of patients, often including also matching samples from healthy tissues. The database includes data from gene expression profiling, copy number variations, SNP genotyping, genome-wide DNA methylation, microRNA profiling, and exon sequencing. Similarly, the UK-based Catalogue of Somatic Mutations in Cancer (COSMIC `http://cancer.sanger.ac.uk/cosmic/`) is an online database of somatically acquired mutations found in human cancers. The project collects data from the scientific literature and from large-scale experiments conducted by the Cancer Genome Project at the Sanger Institute. Furthermore, the International Cancer Genome Consortium (ICGC) coordinates large-scale cancer genome studies in tumors and maintains a database containing genomic, transcriptomic and epigenomic changes from 50 different tumor types (`https://icgc.org/`). Finally, scientific journals and funding agencies increasingly require that raw data underlying published research should be made available in public repositories such as the Gene Expression Omnibus (GEO `http://www.ncbi.nlm.nih.gov/geo/`) for transcriptome data.

Despite the growing amount of data that is now available to researchers, using the data to understand cancer initiation and progression has proven difficult. The main problem stems from the heterogeneity of the data, an issue arising from intrinsic and extrinsic factors. Intrinsic problems come from the fact that mutation patterns vary consistently from patient to patient, even in presence of the same tumor. This happens not only because cancer typically results from several mutations, but also because similar cancer phenotypes can occur through different mutations. Furthermore, most mutations are not tumorigenic but are just passenger mutations that, however, complicate the analysis. Similarly, gene expression profiles are intrinsically noisy even in cells coming from the same patient. Hence, finding cancer signatures in this vastly heterogeneous landscape is a formidable challenge. To the intrinsic problems, we should add extrinsic problems in the data themselves. Searching for signatures relies on the comparison between genetic or expression profiles from different samples and possibly from different studies. Yet, there is often variability introduced by the experimental methods or the sequencing platforms used in each study. This creates problems when collecting and comparing data stemming from different groups and consortia.

Due to the problems discussed above, searching for signatures of cancer in genomic data is still problematic, and one often has to confine the study to homogeneous data sets where extrinsic systematic noise has been reduced as much as possible. This in practice restricts the number of cases that can be analyzed for each particular tumor to a few hundred, at best. While this number may seem large enough, it is often not, especially when we compare it with the size of the human genome, estimated at 20–25,000 genes. Searching for variations in the expression of 20,000 genes from 100 cases is a daunting task since typical statistical clustering

methods fail when the number of samples is much smaller than the dimensionality
of the sample. One interesting way to overcome this problem has recently been put
forward by the group of Eytan Domany (Drier et al., 2013; Livshits et al., 2015).
The key idea is to shift the attention from genes to pathways, comparing how the
expression levels of the genes in each pathway differ in tumor and healthy tissues.
This method effectively reduces the dimensionality of the sample from over 20,000
genes to roughly a few hundred genes present in each pathway, a number that is
now equal to or smaller than the number of available samples.

2.5 Cancer Stem Cells

Due to the accumulation of driver and passenger mutations, cancer cells form an
heterogeneous population. The conventional view was that, despite this wide heterogeneity, all the cells in a tumor were equally tumorigenic. A new idea emerged,
however, two decades ago stating that cancer cells are organized hierarchically,
with cancer stem cells (CSCs) at the top of the hierarchy (see Figure 2.6) (Bonnet and Dick, 1997). According to the CSC hypothesis, tumors behave in analogy
with normal tissues, whose growth is controlled by a small population of slowly
replicating stem cells with the dual capacity of self-renewal and differentiation
into the more mature cells required by the tissue. The role of CSCs has important
implications for therapeutic approaches. According to the conventional view, the
success of a treatment is measured by the number of tumor cells killed; in contrast,
according to the CSC hypothesis, only the CSC subpopulation matters in the end
for complete eradication.

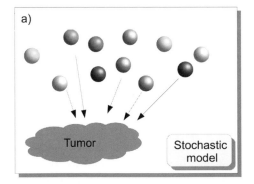 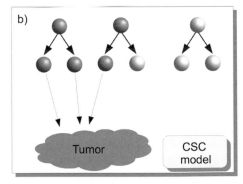

Figure 2.6 a) The stochastic model states that cancer cells are heterogeneous but
all of them are tumorigenic. b) According to the CSC hypothesis, only a subset
of cells are tumorigenic. CSCs can divide symmetrically, giving rise to two CSCs
or asymmetrically giving rise to one CSC and one more differentiated cancer cell.
Those cells are not tumorigenic and cannot produce CSCs.

First evidence for the existence of CSCs came from hematological tumors (Bonnet and Dick, 1997) and later from solid tumors such as breast cancer (Cho et al., 2008) and melanoma (Monzani et al., 2007; Klein et al., 2007; Hadnagy et al., 2006; Schatton et al., 2008; Keshet et al., 2008; Boiko et al., 2010; Roesch et al., 2010; Taghizadeh et al., 2010). CSCs are usually defined by two minimal properties: the first one is the capability to re-grow the tumor from which they were isolated or identified, which implies that the tumor-initiating cells can only be defined experimentally *in vivo*. The second property is the multipotency of lineage differentiation. Thus, to identify CSCs experimentally, the strategy is to isolate cancer cells according to specific molecular markers, usually characteristic of tissue stem cells. The isolated cells are then transplanted into immuno-compromised mice (a tumor xeno-graft). If they give rise to a tumor, the whole process of isolation and transplantation is repeated with new cells obtained from the new tumor (serial transplantation). Alternatively, CSCs are isolated from a mouse tumor and serially transplanted in mice (syngenic transplantation). Both xeno- and syngeneic transplantations might, however, misrepresent the real network of interactions with diverse support such as fibroblasts, endothelial cells, macrophages, mesenchymal stem cells and many of the cytokines and receptors involved in these interactions (for a more comprehensive discussion read La Porta and Zapperi, 2013). As a consequence of this, contrasting results have appeared in the literature for the frequency of CSCs in xenograft tumors. Despite these limitations, recent experiments *in vivo* have confirmed the presence of an aggressive CSC-like subpopulation in benign and malignant tumors (Schepers et al., 2012; Driessens et al., 2012; Chen et al., 2012). The population dynamics of CSCs is, however, still debated: a recent paper points out the possibility that non-CSCs breast cancer cells can revert to a stem cell-like state even in the absence of mutations (Gupta et al., 2011). The way this type of phenotypic switching occurs was recently elucidated in melanoma, where CSC-like cells are dynamically regulated by a complex miRNA network (Sellerio et al., 2015).

2.6 Feeding the Tumor: Angiogenesis

Tumors induce new blood vessels from pre-existing ones to receive nutrient and oxygen and to evacuate metabolic waste and carbon dioxide. This process is called *angiogenesis* and it was proposed as a hallmark of cancer (Hanahan and Weinberg, 2011). During physiological conditions such as wound healing and female reproductive cycling, angiogenesis is turned on, but only transiently. In contrast, during tumor progression, an angiogenic switch is almost always activated and remains on, causing normally quiescent vasculature to continually sprout new vessels (Hanahan and Folkman, 1996). Angiogenesis switches on or off depending on a balance between positive and inhibitory factors (Baeriswyl and Christofori, 2009). There is an initial triggering of the angiogenic switch during tumor development that is followed by a variable intensity of ongoing neovascularization, controlled by a complex biological process involving both the cancer cells and the associated stromal microenvironment (Baeriswyl and Christofori, 2009). In this connection, pericytes, a specialized mesenchymal cell type that wraps around the endothelial tubing of blood vessels, are becoming the object of increasing interest. In normal tissues, pericytes are known to provide paracrine support signals to the normally quiescent endothelium. They also collaborate with the endothelial cells to synthesize the vascular basement membrane that anchors both pericytes and endothelial cells and helps vessel walls to withstand the hydrostatic pressure due to blood flow. The mainstream idea was that tumor-associated vasculature lacked appreciable coverage by these auxiliary cells. Careful microscopic studies conducted in recent years have revealed, however, that pericytes are associated with the neovasculature of most, if not all, tumors (Raza et al., 2010).

The process leading to vascular sprouting is dynamic and involves distinct types of endothelial cells. A tip cell that bears extending filipodia moves at the front of a sprouting vessel and is followed by proliferating stalk cells elongating the sprout (Carmeliet and Jain, 2011). VEGF activates tip cells, whereas the signalling protein Notch suppresses them and stimulates stalk cells. Interestingly, glycolytic enzymes are selectively compartmentalized within tip cell filipodia (De Bock et al., 2013). The same paper suggested that tip cell protrusions extending into areas of low oxygen tension are able to generate local ATP for processes requiring high energy, such as the assembly of cytoskeletal actin filaments (De Bock et al., 2013).

2.7 Metastasis

Malignant cancer cells can spread through organisms using the blood or lymphatic system (see Figure 2.7). The metastatic process is considered an inefficient multistep process. During metastasis, cancer cells should overcome several obstacles when they move through the tissues and the extracellular matrix, which can exhibit different topology and mechanical stiffness, enter or exit from lynphatic vessels and resist shear and compressive stresses. To perform these tasks, cancer cells undergo transitions that perturb basic cell processes, such as the expression of new surface receptors and changes in their cytoskeleton or in their directional polarity. It is generally believed that the traits needed by cancer cells to survive the long travel to distant organs are acquired before the metastatic process starts. This idea came from the identification in primary tumors of gene expression signatures characteristic of metastatic cells (Vanharanta et al., 2013; Zhang et al., 2013). It is likely, however, that biophysical forces also play a role in the process.

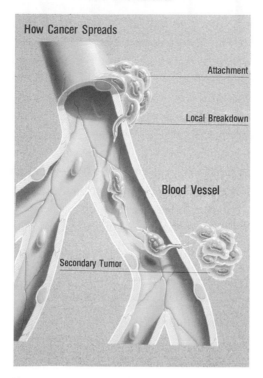

Figure 2.7 Spreading of cancer cells. Metastatic cells attach to the basement membrane, then infiltrate and move through the blood stream, and eventually seed a secondary tumor. Image by Jane Hurd, National Cancer Institute, public domain.

A likely site for selection of metastatic traits in primary tumors is at the invasive front, the intersection between an advancing tumor mass and the surrounding stroma. Cancer cells that have overcome their own oncogenic stresses, the counteracting response of the cell to oncogene activation, can become invasive. The invasion process exposes the cell to further stress due to the reactive stroma, surveillance from the immune system and lack of oxygen (hypoxia). Genetics is not the only important factor for metastasis. Recently large-scale genome sequencing studies have shown more similarities than differences between primary tumors and their metastases (Yachida et al., 2010).

The organs affected by metastasis depend on the probability that metastatic cancer cells reach distant organs, but also survive and grow there, initiating further metastasis. Disseminated cancer cells need supportive sites to establish a metastasis, in a way that is reminiscent of stem cell niches. For stem cells, the location and constitution of stem cell niches has been defined in various tissues, including the intestinal epithelium, hematopoietic bone marrow, epidermis and brain. In a similar way, the term *metastatic niche* is used to designate the specific locations, stromal cell types, diffusible signals and ECM proteins that support the survival and self renewal of disseminated metastatic cells. Metastatic cells could occupy a stem cell niche, including perivascular sites recruiting stromal cells that produce stem cell niche-like components or by producing these factors themselves. Some tumors release systemic suppressor factors that make micrometastasis dormant, but others, such as melanoma, erupt decades after a primary tumor has been eradicated surgically or pharmacologically, possibly due to a reactivation of dormant micrometastasis. Recent papers show that nutrient starvation can induce autophagy, causing the cancer to shrink and adopt a reversible dormancy. When the tissue changes, becoming more accessible to nutrients, the tumor resumes its growth (Kenific et al., 2010). To conclude, this relationship between genetic factors and microenvironment is still debated: it seems plausible that both factors contribute to metastasis.

2.8 Diagnostic Methods

The main goal of diagnostic methods is to identify the presence of cancer cells as soon as possible. This goal is, however, very difficult to achieve, since usually people go to the doctor when they have some symptoms, meaning that the tumor has already grown significantly. When diagnosed, a primary tumor may have already seeded distant organs with cancer cells, including cells with tumor-initiating capacity. After diagnosis, the primary tumor may be removed by surgery and irradiation, and disseminated cancer cells may be eliminated by systemic chemotherapy, leading to a cure. Alternatively, residual metastatic cancer cells may remain in a latent state, eventually giving rise to new metastasis. New rounds of therapy may then induce regression of the metastatic lesions, but chemoresistant metastatic cancer cells selected during each round of treatment may eventually yield uncontrollable metastasis. This process is responsible for 90% of deaths due to cancer.

Two main goals have been the subject of intensive investigations in the last ten years. First, the search for biomarkers that would allow the prevention of the dissemination of the cancer as early as possible, mainly in selected sub-populations of people with a given age or familiar history. Secondly, the identification of biomarkers that would help follow the effect of treatment (chemo- or radio-therapy). From the theoretical point of view, this kind of approach tries to identify key aspects of the tumor such as its proliferative capacity, or signatures for angiogenesis or metastasis. On the other hand, the huge literature that came out in recent years shows that the identification of possible bio-markers is often specific to distinct tumors and not always reproducible for other tumors or even easily applied to all the patients. Moreover, the evidence for a hierarchic organization of the tumor and, more recently, the possibility for a cell to switch from cancer cell to CSC (Gupta et al., 2011; Sellerio et al., 2015) further complicates our general view of tumor heterogeneity and therefore possible treatment and identification strategies. In particular, Sellerio et al. (2015) clearly show that eradicating CSCs might not be a viable strategy to treat tumors, suggesting instead to target the mechanisms by which cancer cells switch back to the CSC state.

3

A Modeling Toolbox for Cancer Growth

In this chapter we review a set of basic mathematical models and tools that have been widely applied to study cancer growth. We start in Section 3.1 with branching processes, simple probabilistic mean-field models for the evolution of a population. Branching processes represent the most widely used approach to model population dynamics of cancer cells. In Section 3.2 we consider probabilistic models for the evolution of mutations in cancer development. These models are sometimes related to branching processes we need to take into account for mutations in expanding clonal populations. In Section 3.3 we discuss models' gene regulatory networks and signaling networks relevant for cancer, and in particular models for the p53 network. Mean-field models are not adequate to take into account the spatial localization of a tumor or of cancer cell population. To this end, one should introduce individual cell models in which we can follow the dynamics of a set of interacting active cells (Section 3.4). At a more coarse-grained level, the growth dynamics of a tumor can be represented by lattice cellular automata or by continuum phase-field models as discussed in Section 3.5.

3.1 Branching Processes

Branching processes (Harris, 1989; Kimmel and Axelrod, 2002) are a class of simple models that have been used extensively to model growth dynamics of stem cells (Vogel et al., 1968; Matioli et al., 1970; Potten and Morris, 1988; Clayton et al., 2007; Antal and Krapivsky, 2010; Itzkovitz et al., 2012) and cancer cells (Kimmel and Axelrod, 1991; Michor et al., 2005; Ashkenazi et al., 2008; Michor, 2008; Tomasetti and Levy, 2010; La Porta et al., 2012). Branching processes can be defined in discrete or continuous time and with evolution rules that may or may not depend on time. A detailed review of the mathematical theory of the branching process is given by Harris (1989) and applications to biology are discussed in Kimmel and Axelrod (2002).

3.1.1 The Galton–Watson Model

The first example of a branching process was introduced by Galton and Watson in the second half of the nineteenth century to study the extinction of family names. The Galton and Watson model is a probabilistic model for the evolution of a population defined at discrete times $k = 0, 1, \ldots$ Let us denote by X_k the number of individuals in the population at time k. In the context of cancer, we can think of X_k as the number of cells in a tumor or in a colony in vitro at time $t_k = kt_0$, where t_0 is the typical division time of the cell (see Figure 3.1). In the following we assume that $X_0 = 1$, or that the cell population starts from a single clone. The branching process is defined by a set of probabilities p_i that a single individual is replaced by $i = 0, 1, 2, \ldots, N$ individuals at the next time step, assuming that p_i does not depend on time and that individuals reproduce independently. Normalization of probabilities imposes that

$$\sum_{i=0}^{N} p_i = 1. \tag{3.1}$$

In the case of cells, we can set $N = 2$ so that a cell can die with probability p_0, remain quiescent with probability p_1 and divide with probability p_2.

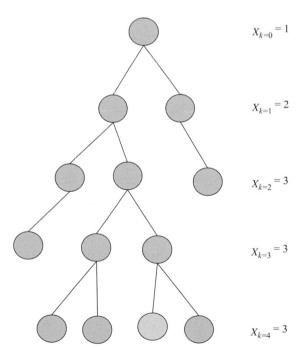

$X_{k=0} = 1$

$X_{k=1} = 2$

$X_{k=2} = 3$

$X_{k=3} = 3$

$X_{k=4} = 3$

Figure 3.1 An illustration of Galton–Watson branching process.

Since the state of the system at time k only depends on the state at time $k - 1$, the branching process is a Markov process defined by transition probabilities $P_{ij} \equiv P(X_{k+1} = j | X_k = i)$. Once the process is defined, we are interested in computing some measures of the evolution of the population such as the probability distribution of having n individuals or cells at time k or $P(X_k = n)$, or the average number of individuals at time k, or the probability that the population is extinct.

If we are interested in modeling the growth curve of a cancer cell population or of a tumor, we can write a simple recursion relation for the average number of cells $\mu_k \equiv \langle X_k \rangle$ after k generations:

$$\mu_k = \sum_{j=0}^{N} j p_j \mu_{k-1}. \tag{3.2}$$

Equation 3.2 can be derived by noticing that each cell evolves independently according to the probabilities p_j. Since each cell can only divide into two ($N = 2$), we can write Equation 3.2 as

$$\mu_k = (p_1 + 2p_2)\mu_{k-1}, \tag{3.3}$$

yielding $\mu_k = (p_1 + 2p_2)^k \mu_0$.

The first important observation is that the population grows exponentially as long as the branching ratio is less than one, or $r = \mu_k / \mu_{k-1} = (p_1 + 2p_2) > 1$, while it decreases for $r = \mu_k / \mu_{k-1} = (p_1 + 2p_2) < 1$. These two cases are denoted as supercritical and subcritical, respectively. In the critical branching process ($r = \mu_k / \mu_{k-1} = 1$) the average size of the population remains constant.

To make contact with experiments, we can write $k = t/t_0$ and define the cumulative population doubling (*CPD*) of a cell population as the number of times the population doubles its size:

$$CPD \equiv \frac{\log(\mu_k / \mu_0)}{\log(2)} = \frac{t \log(p_1 + 2p_2)}{t_0 \log(2)}. \tag{3.4}$$

Equation 3.4 shows that CPD grows linearly in time as $CPD = R_d t$, and can be used to fit experimental curves of in vitro cancer cell growth (Figure 3.2). The growth rate $R_d = \log(p_1 + 2p_2)/(t_0 \log(2))$ can then be interpreted in terms of the branching process model.

3.1.2 Generating Functions

A useful function to compute several statistical properties of the branching process is the generating function of X_k defined as

$$f^{(k)}(s) \equiv \sum_{j=1}^{\infty} P(X_k = j)s^j. \tag{3.5}$$

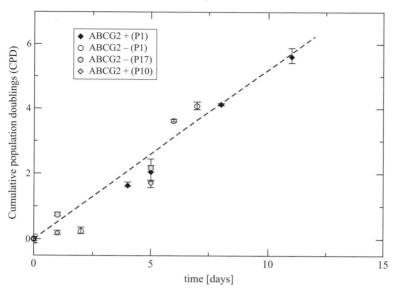

Figure 3.2 In vitro growth curves for melanoma cell populations can be interpreted using the Galton–Watson model. The experimental curves are obtained melanoma (IgR39) cell lines that have been sorted by flow cytometry according to the expression of a marker (ABCG2). The experiments were performed at different times (or passages P1, P10 or P17, marking the number of times cells have been split and transferred from one dish to another). In all cases, the CPD grows linearly with a slope $R_d \simeq 0.52$ with only small variations between the different experiments.

Since $X_0 = 1$, we have that $f^{(0)} = s$. We also notice that, by definition, the first iteration of the generation function is equal to the generation function for the branching process probabilities $f(s)$, or

$$f^{(1)}(s) = f(s) \equiv \sum_{j=1}^{\infty} p_j s^j. \tag{3.6}$$

In the simple branching process describing cancer cells, the probability generating function can be simply evaluated as $f(s) = p_1 s + p_2 s^2$.

The generating function represents in a compact form all the moments of the distribution $P(X_k)$ (Harris, 1989; Kimmel and Axelrod, 2002). In particular, the first moment is given by

$$\mu_k = \frac{df^{(k)}(s)}{ds}\bigg|_{s \to 0}, \tag{3.7}$$

and the second moment by

$$\langle X_k^2 \rangle = \frac{d^2 f^{(k)}(s)}{ds^2}\bigg|_{s \to 0} + \frac{df^{(k)}(s)}{ds}\bigg|_{s \to 0}. \tag{3.8}$$

Due to the tree structure of the Galton–Watson model, the generating function obeys a recursion relation (Harris, 1989):

$$f^{(k+1)}(s) = f(f^{(k)}(s)) \tag{3.9}$$

It is straightforward now to use Equation 3.9 to derive the recursion relations for the moments such as Equation 3.2. In the case of the simple branching process for cells, Equation 3.9 takes the simpler form:

$$f^{(k+1)}(s) = p_0 + p_1 f^{(k)}(s) + p_2 [f^{(k)}(s)]^2. \tag{3.10}$$

One can also use Equation 3.9, to obtain the evolution of the generating function and the asymptotic properties of $P(X_k)$. In this respect, a very interesting result is the asymptotic probability of extinction (i.e., the probability that $X_k = 0$ after a finite number of generations k):

$$q \equiv P(X_k \to 0) = \lim_{k \to \infty} f^{(k)}(0). \tag{3.11}$$

As demonstrated in Harris (1989), $q = 1$ for critical and subcritical branching processes, while it is equal to the unique nonnegative solution of $q = f(q)$ for supercritical branching processes. In the case of the simple model for cells, the solution can be obtained explicitly as

$$q = \frac{1 - p_1}{2p_2} - \frac{\sqrt{(1 - p_1)^2 - 4p_0 p_2}}{2p_2}. \tag{3.12}$$

The intriguing aspect of this result is that, even if the branching process is super-critical, there is a finite probability that it will become extinct after some time. Furthermore, critical branching processes in which the population remains constant on average, in practice become extinct for sure. Extinction in this case is caused by fluctuations: if we consider the evolution of a single clone, sooner or later a fluctuation will determine its extinction.

3.1.3 Continuous Time Branching Processes

The Galton–Watson branching process assumes discrete time, so that cells reproduce at fixed time intervals. If we want to take into account the variability in division time, we should turn to continuous time branching processes (Harris, 1989). The natural generalization of the Galton–Watson model for cells is the birth-and-death model where each cell has a probability $\beta(t)dt$ of dying in the time interval $(t, t + dt)$ and a probability $\gamma(t)dt$ of dividing into two cells in the same time interval. Here the rates of cell death and division have been written as time dependent, but for simplicity we can assume they are constant.

The birth-and-death process can be used to compute the evolution of the size distribution of cancer cell colonies $p(s, t)$, defined as the probability that a single cancer cancer cell gives rise to a colony composed by s cells at time t, where time is measured in days. The probability density function $p(s, t)$ evolves according to the following master equation

$$\frac{dp(s, t)}{dt} = \gamma(s - 1) \, p(s - 1, t) + \beta(s + 1) \, p(s + 1, t) - (\gamma + \beta)s \, p(s, t), \quad (3.13)$$

starting with an initial condition $p(s, 0) = \delta_{s,1}$.

From Equation 3.13, we can obtain an equation for the first moment, the average colony size $\langle s \rangle$,

$$\frac{d \langle s \rangle}{dt} = (\gamma - \beta) \langle s \rangle, \quad (3.14)$$

yielding an exponential growth

$$\langle s(t) \rangle = \exp[(\gamma - \beta)t]. \quad (3.15)$$

Here, we find again the distinction between subcritical branching processes for $\gamma < \beta$, when the colony disappears in the long-time limit, supercritical branching processes for $\gamma > \beta$ where the colony grows exponentially and finally critical branching process where the colony size is constant.

An explicit solution of Equation 3.13 can also be obtained (see Harris 1989, page 104):

$$p(0, t) = 1 - \frac{\langle s(t) \rangle (\gamma - \beta)}{\gamma \langle s(t) \rangle - \beta} \quad (3.16)$$

$$p(s, t) = \langle s(t) \rangle \left(\frac{\gamma - \beta}{\gamma \langle s(t) \rangle - \beta} \right)^2 \left(1 - \frac{\gamma - \beta}{\gamma \langle s(t) \rangle - \beta} \right)^{s-1} \quad \text{for } s \geq 1. \, (3.17)$$

In the limit $\beta \to 0$, the model is equivalent to the Yule problem (Harris, 1989) and the solution is given in simpler form by (Harris, 1989):

$$p(s, t) = e^{-\gamma t}(1 - e^{-\gamma t})^{s-1} \quad \text{for } s \geq 1. \quad (3.18)$$

The theoretical results derived above can be compared with typical cancer cell colony growth experiments, such as in the crystal violet assay, where N_0 cells are plated on multiwells and stained after a fixed number of days (e.g., 8 or 10 days) (Baraldi et al., 2013). Typically, in this type of experiment, researchers are only interested in the number of colony-forming clones. The experiments, however, provide a wealth of other useful information concerning the distributions of colony sizes and shapes that are rarely analyzed in detail (Baraldi et al., 2013). This is illustrated by Figure 3.3 displaying the cumulative distribution of colony sizes grown for 8 or 10 days, collecting together data obtained in different wells

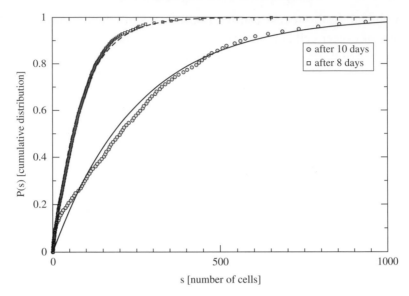

Figure 3.3 The distribution of colony sizes after 8 and 10 days of growth. The distribution is obtained collecting all the colonies obtained for all the different values of N_0, corresponding to 32 and 30 wells, 2347 and 1067 colonies of average size $\langle S \rangle \simeq 85$ and $\langle S \rangle \simeq 258$ for 8 and 10 days respectively. The curves represent the simulation of the continuum time branching process using the parameters obtained from the maximum likelihood estimate.

and for different values of N_0. The cumulative distribution $P(s)$ is related to the probability density function $p(s)$ by

$$P(s) = \sum_{s'=1}^{s} p(s').$$

(3.19)

These results were obtained using semi-automatic image analysis methods to quantify the size of each colony, allowing us to study with relative ease thousands of colonies formed by hundreds of cells. As a comparison, earlier studies based on manual counting of the cells in the colonies could perform statistics over around 50 colonies (Kimmel and Axelrod, 1991).

To obtain a quantitative comparison between theory and experiments, one can employ the maximum likelihood method and estimate the best values for γ and β, with the constraints $\beta \geq 0$ and $\gamma \geq 0$. In practice, using an iterative optimization scheme one finds the values of γ and β that maximize the cost function given by

$$\mathcal{L} = \sum_{i} \log(p(s_i^{(t=8)}, 8)) + \sum_{j} \log p(s_j^{(t=10)}, 10),$$

(3.20)

where $s_i^{(t=8)}$ and $s_j^{(t=10)}$ are the experimentally measured colony sizes after 8 and 10 days, respectively. Using this scheme, the best fit yields $\gamma = 0.55$ divisions/day and sets β to its limiting value of zero. The fitted value for γ can be compared with previous experiments on melanoma growth (see Figure 3.2), but at much higher cell density, which yields $\gamma \simeq 0.4$ divisions/day (La Porta et al., 2012). This is compatible with the general idea that the growth rate decreases when the cell density is higher.

3.2 Mutation Models of Cancer

Mutations in Homeostatic Cell Populations

In Chapter 2, we briefly discussed how oncogenes and tumor suppressor genes can give rise to cancer as a consequence of a single mutation in one allele or from two mutations in both alleles, respectively. This results in distinct predictions for the probability that tumors are first diagnosed at a given time, as illustrated in Figure 2.3 (Knudson, 1971). In this section, we provide more details on how these results arise from simple probabilistic models of mutations (see Nowak 2006 for a pedagogical review).

Tissues can be subdivided into compartments composed by dividing cells in homeostasis, so that their number N is held approximately constant. In a homeostatic compartment, for each cell that divides there is on average another cell that dies so that the number of cells is kept in equilibrium. A typical example of a cell compartment is the colonic crypt, where stem cells at the base of the crypt give rise to more differentiated cells that divide and eventually leave the crypt or die (Figure 3.4). In the language of branching processes, a homeostatic compartment corresponds to a critical branching ratio, where $r - 1$ and the average number of offspring of each cell is exactly one. Cancer represents a breakdown of homeostasis since $r > 1$ and cells proliferate uncontrollably.

To study the emergence of deleterious mutations in oncogenes and tumor suppressor genes, we consider a cell compartment of size N and compute the probability that one or two mutations arise in one cell. Consider first the case of oncogenes, where we are interested in the time at which the first deleterious mutation arises in the cell population. If the mutation rate per gene per cell division is R, the probability that at least one mutation occurs in the population at time t, measured in units of cell cycles, is simply

$$P(t) = 1 - \exp(-NRt). \tag{3.21}$$

We recognize here the expression used by Knudson to fit single hit data, since the probability that no mutation arises at time t is just $1 - P(t)$.

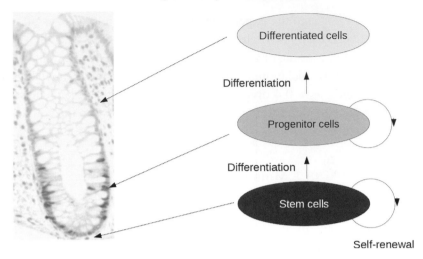

Figure 3.4 An image of a colonic crypt, composed by stem cells at the base that self-renew and give rise to progenitor cells. Those cells in turn give rise to differentiated cells. The process is in homeostatic equilibrium when the total number of cells removed from the crypt is equal to the number of newly generated cells. Image by Bravo and Axelrod, 2013, CC-BY-2.0 licence.

3.2.1 The Moran Process

One can extend this calculation, leading to Equation 3.21, to take into account also the probability not only that the mutation occurs but also that it eventually will spread through the entire compartment. This process can be described mathematically by the Moran model (Moran, 1958), a particular birth-and-death process in which a fixed-size population is composed of two species of cells, A and B (in our case, mutated and non-mutated cells), that compete for dominance. At each time step of the process we choose a cell at random and replace it with another cell which can belong to the same species or to the other. This operation encapsulates two classes of events that could happen in the population: a cell of type A/B divides and one cell of the same type dies, or a cell of type A/B divides and a cell of a different type dies. The first class of events does not change the relative composition of the population, while the second one does (see Figure 3.5).

The state of the population can be defined by the number i of cells of type A, where the total number of cells is denoted by N. At each time step the state, the probability to pick a cell of type A, is i/N and a cell of type B is $(N - i)/N$. We can then compute transition probabilities from the state i to states $i \pm 1$ as

$$P_{i,i\pm1} = \frac{i(N - i)}{N^2} \tag{3.22}$$

$$P_{i,i} = 1 - P_{i,i+1} - P_{i,i-1}. \tag{3.23}$$

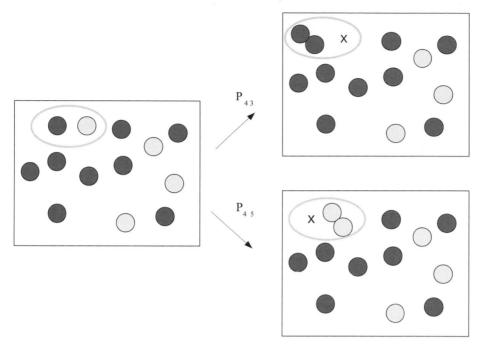

Figure 3.5 The Moran model. Cells of two types (pale and dark) coexist. The number of pale cells here is $i = 4$. At the next time step two possible changes can occur: (top) a pale cell dies (marked by x) and a dark cell divides, decreasing the number of pale cells by one; or (bottom) a dark cell dies (marked by x) and a pale cell divides, increasing the number of pale cells by one. If cells of the same type divide and die, the configuration remains the same.

Since $P_{0,0} = P_{N,N} = 1$, the states with all cells of type A or B are absorbing states of the process. Hence, a mixed population will eventually mutate into one of the two species. The probability to end up in state 0, starting from state i, is simply given by $x_i = i/N$.

We can further complicate the problem by introducing a fitness f for cells of type A, while cells of type B have fitness 1. If $f > 1$, cells of type A have a selective advantage over those of type B. This is reflected in the evolution equations that now read

$$P_{i,i+1} = \frac{fi(N-i)}{(fi+N-i)N} \tag{3.24}$$

$$P_{i,i+1} = \frac{i(N-i)}{(fi+N-i)N} \tag{3.25}$$

$$P_{i,i} = 1 - P_{i,i+1} - P_{i,i-1}. \tag{3.26}$$

In this case, the probability to end up in the absorbing state A starting from state i is given by

$$x_i = \frac{1 - f^i}{1 - f^N}, \qquad (3.27)$$

which in the limit of $f \to 1$ reduces to the previous expression. Equation 3.27 allows the probability that a random mutation in a single cell takes over the entire population to be computed. This "fixation" probability is given by $\rho \equiv x_1 = (1 - f)/(1 - f^N)$.

Coming back to the one-hit mutation model, we can generalize Equation 3.21 to take into account the fitness advantage of mutations and their probability of fixation ρ. This is important because deleterious mutations could be eliminated, not leading to a tumor, unless they are fixated in the compartment. Hence, the one-hit probability in Equation 3.21 takes the form

$$P(t) = 1 - \exp(-NR\rho t). \qquad (3.28)$$

3.2.2 Multiple-Hit Mutation Models

Equations 3.21 and 3.28 describe the probability that a mutation in an oncogene occurs and becomes fixated at time t. As discussed in Chapter 2, to describe cancer-initiating mutations in oncosuppressor genes we need to consider two mutations, occurring in both alleles. The calculation in this case is much more complicated, and a simple closed-form solution can be obtained only in specific limits (Nowak, 2006). The parameters of the problems here are the number of cells N and the mutation rates for the first and second hit, R_1 and R_2, respectively.

Here we discuss only the result in the limit of large populations, $N \gg 1/R_1$, while the derivation for small and intermediate populations can be found in Nowak (2006). In this limit, we can assume that the first mutation occurs almost immediately in at least one cell and we have only to estimate the probability that a second mutation also hits a mutated cell. The number of cells with the first mutation grows linearly as $x_1 = NR_1t$ and the probability that a second mutation occurs at time t is given by

$$P_2(t) = 1 - \exp(-R_2 \int_0^t d\tau x_1(\tau)) = 1 - \exp(-R_2 R_1 N t^2), \qquad (3.29)$$

which is the expression used by Knudson to fit the two-hit curve for the incidence of retinoblastoma (Figure 2.3).

It is possible to generalize these mutation models to more complex situations, considering for instance the growth advantages of each mutation in expanding clonal populations. For instance, Beerenwinkel et al. (2007) proposed a model of clonal expansions in a fixed-sized or slowly growing cell population. In the model, new driving mutations trigger consecutive clonal expansions leading to different clonal waves, corresponding to different levels of malignancy, a property also observed in real tumors (Jones et al., 2008). More recently, Bozic et al. (2010) estimated the number of driver mutations as a function of the total number of mutations in a tumor.

3.3 Simulations of Gene Regulatory Networks

3.3.1 Kinetic Reaction Equations

A very large research activity is devoted to modeling gene regulatory networks and signaling networks relevant for cancer (de Jong, 2002). As discussed in Section 1.6, gene expression is regulated by complex networks involving interactions of thousands of nodes. Mutations in individual genes can lead to cascade effects involving other genes, and it is difficult to anticipate the effect of each perturbation (Serra et al., 2004, 2007). In the most detailed form, regulatory networks and pathways can be modeled by a set of coupled reaction-kinetics equations that in principle could describe in a very accurate way the response of the network (Bhalla and Iyengar, 1999). In general, these models involve reversible reactions of the type

$$A + B \underset{k_f}{\overset{k_b}{\rightleftharpoons}} AB, \tag{3.30}$$

or

$$A + B \underset{k_f}{\overset{k_b}{\rightleftharpoons}} C + D, \tag{3.31}$$

or else irreversible reactions such as

$$A + B \overset{k_i}{\rightarrow} AB. \tag{3.32}$$

Here, k_b, k_f and k_i are suitable reaction rates. The reactions above describe in synthetic form kinetics equations. For instance, Equation 3.30 translates into

$$\frac{d[A]}{dt} = k_b[AB] - k_f[A][B], \tag{3.33}$$

where $[X]$ is the concentration of the species X. Thus, a complex signaling network, like the one representing the Wnt pathway reported in Figure 3.6 (Lee et al., 2003), is reduced to a set of coupled differential equations.

The key problem with the reaction kinetics approach discussed above is that reaction rates for the equations are, most of the time, unknown or known only approximately. Reconstructing the full set of reaction rate parameters in a pathway is indeed a daunting experimental task (Lee et al., 2003). Furthermore, obtaining model parameters from experiments typically involves extensive use of multi-parameter fits. A striking observation is that it is often possible to fit the same

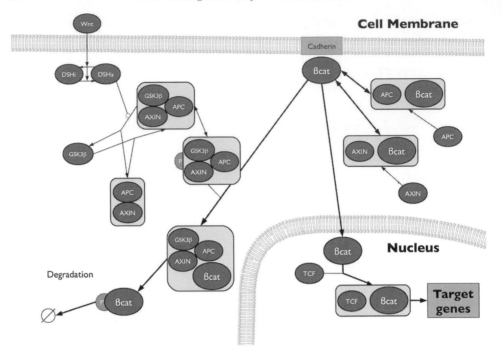

Figure 3.6 A reaction network for the Wnt signaling pathway. The model has been proposed by Lee et al. (2003) and corresponds to a set of coupled differential equations.

set of experimental observations by varying parameter values by orders of magnitude. Sethna and coworkers coined the expression "sloppy model" to describe this kind of situation (Brown et al., 2004; Gutenkunst et al., 2007). In sloppy models, parameters can be varied in extremely large ranges without affecting the final result. Some parameter combinations are, however, stiff, and their variation leads to significant deviations in the results. This observation sometimes allows the network model to be simplified, omitting sloppy variables and focusing instead on the stiff ones (Brown et al., 2004; Gutenkunst et al., 2007).

3.3.2 Kinetic Reaction Model for P53 Oscillations

As a concrete illustration of kinetic reaction models used for gene regulatory networks, we consider the oscillatory dynamics of the expression of p53 (Lev Bar-Or et al., 2000), the "guardian of the genome," a key oncosuppressor discussed in Section 2.3. Here we describe the kinetic reaction model introduced by Tiana et al. (2002) to account for this behavior. As shown in Figure 3.7, the model includes three species p53, with concentration $[p53] = P$, MDM2 with

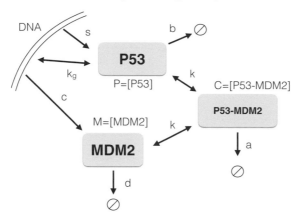

Figure 3.7 A reaction network model for the kinetics of p53 and MDM2. The model has been proposed by Tiana et al. (2002) and corresponds to the coupled differential equations 3.34–3.36.

concentration $[MDM2] = M$ and the p53–MDM2 complex, with concentration $[p53 - MDM2] = C$. The p53 protein is produced with rate s and is degraded with rate b. Since p53 induces the expression of MDM2, the MDM2 protein is produced with a rate proportional to the concentration of the p53–MDM2 complex, which is assumed to be in equilibrium with its components. The free proteins of p53 and MDM2 are also in equilibrium with the p53–MDM2 complex. The kinetics is then described by the set of equations

$$\frac{\partial P}{\partial t} = s - aC(t) - bP(t) \tag{3.34}$$

$$\frac{\partial M}{\partial t} = c\frac{P(t-\tau) - C(t-\tau)}{k_g + P(t-\tau) - C(t-\tau)} \quad dM \tag{3.35}$$

$$C = \frac{1}{2}(P + M + k - \sqrt{(P + M + k)^2 - 4PM}). \tag{3.36}$$

Notice that the second equation contains a delay τ in the production of MDM2, reflecting the transcriptional delay after binding with p53. This delay is crucial to observe undamped oscillations, that are instead not observed for $\tau = 0$ (Tiana et al., 2002). This is summarized in Figure 3.8, showing the result of numerical simulations with $\tau = 0$ (top, no oscillations) and $\tau > 0$ (bottom, with clear oscillations).

3.3.3 Boolean Networks

To overcome the intricacies associated with unknown or poorly known reaction rate parameters, one can resort to simpler Boolean models where the nodes in the

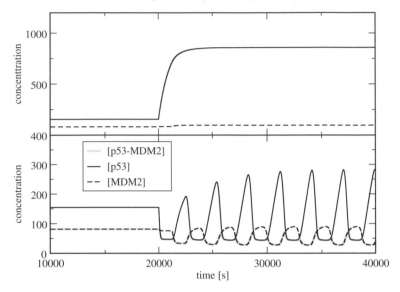

Figure 3.8 The solution of equations 3.34–3.36 for $\tau = 0$ (top) and $\tau > 0$ (bottom). Image by G. Tiana et al., 2002.

pathway are associated with binary values ($s_i = 0, 1$), indicating if the gene is expressed or not, or more broadly if it is under-expressed or over-expressed (Kauffman et al., 2003; Bornholdt, 2008). In these models, reaction equations are replaced by Boolean functions that are evolved according to specific rules, such as random sequential update or synchronous parallel update. Several choices are possible for the Boolean functions. As an illustration, consider threshold networks (Bornholdt, 2008) where the state of a node depends only on the sum of its input signals and the state of a variable $s_i(t)$ evolves in parallel according to

$$s_i(t+1) = \theta(\sum_j w_{ij} s_j(t) + h_i),\qquad(3.37)$$

where $\theta(x)$ is the Heaviside step function, h_i is a threshold parameter and $w_{ij} \pm 1$ represents a set of repressing (-1) and activating ($+1$) links. The structure of these models is similar to that of Ising spin glass models, widely studied in statistical mechanics.

Boolean networks have been widely used to study the regulation of cancer pathways (Zhang et al., 2008b; Steinway et al., 2014; Fumia and Martins, 2013). For instance, Fumia and Martins (2013) constructed a model integrating the main signaling pathways involved in cancer based on known protein–protein interaction networks. Their model illustrates that malignancies could result through different mutation patterns and allows the testing of therapeutic strategies *in silico* (Fumia and Martins, 2013). Other models focus on the epithelial-to-mesenchymal transition in hepatocellular carcinoma (Steinway et al., 2014) or on T cell large granular lymphocyte leukemia (Zhang et al., 2008b).

3.4 Individual Cell-Based Model

All the models discussed in previous sections fail to consider the geometrical arrangement of cancer cells and their physical interactions. Hence, these models do not provide any information on the interaction between cells and their localization inside the tumor or the colony. To overcome this limitation, one can consider individual cell models, where each cell is treated as an elastic particle with suitable rules for its division and motion. Several computational models based on individual cell mechanics have been introduced in the past to simulate cancer growth *in vitro* and *in vivo* (Walker et al., 2004; Holcombe et al., 2012; Galle et al., 2005; Drasdo and Hoehme, 2005, 2012; Baraldi et al., 2013; Taloni et al., 2014).

In the simplest case, cells interact with a simple repulsive Hertz pair potential (Drasdo et al., 1995):

$$V(r_{ij}) = \frac{4E(2R - r_{ij})}{15(1 - v^2)} \sqrt{\frac{R}{2}}, \text{ for } r_{ij} < 2R, \tag{3.38}$$

where R is the cell radius, r_{ij} is the distance between the cell centers, E is the Young modulus and v is the Poisson ratio (see Section 6.1). One can then include suitable adhesive interactions between neighboring cells by approximate potentials (Chu et al., 2005; Drasdo and Hoehme, 2005, 2012) or by introducing elastic links between cells (Taloni et al., 2014).

Cell division can occur simply randomly at rate γ. To this end, one can use the Gillespie algorithm (Gillespie, 1976), randomly selecting a cell and performing a cell division. The next time for each division is chosen according to a Poisson distribution with rate $N\gamma$, where N is the number of cells at time t. An alternative is to couple the division rate to the mechanical state of the cell, so that compressed cells divide more rarely (Taloni et al., 2014). Cell division is simulated by replacing the dividing cell by two new randomly oriented cells placed at close distance r_0, a dumbbell (Drasdo and Hoehme, 2005). The distance between the cells in the dumbbell is then slowly increased until it reaches R and the the dumbbell is replaced by two new cells (Figure 3.9). During the division process the dumbbell applies forces to neighboring cells and can lead to displacements and rearrangements. Finally, one can consider cell motion either by including a random force giving rise to Brownian motion or by also allowing for active forces leading to cell migration (Sepúlveda et al., 2013).

Individual cell models have been successfully used to model the growth kinetics of cancer cell colonies in two dimensions (Walker et al., 2004; Holcombe

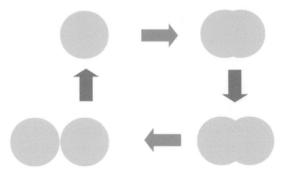

Figure 3.9 The cell cycle in individual cell model. Cells are picked randomly to undergo mitosis. Dividing cells are first replaced by a dumbbell which then expands until two new cells are formed. The cycle is then repeated.

Figure 3.10 A growing nevus simulated with an individual cell model. An example of a configuration of a nevus growing in the epidermis. Cells in the epidermis and the dermis are connected by springs to simulate the ECM. The nevi cells are pushing on the basement membrane as they spread through the tissue. Image by Taloni et al., 2014, CC-BY-4.0 license.

et al., 2012; Galle et al., 2005; Drasdo and Hoehme, 2005, 2012), tumor spheroids in three dimensions (Montel et al., 2012) and the formation of nevi in the skin (see Figure 3.10; Taloni et al., 2014). They provide the basis for more complex models, including random mutations (Drasdo et al., 2007) or different cell types (Taloni et al., 2015b), or for modeling collective cell migration and tissue invasion (Sepúlveda et al., 2013).

3.5 Cellular Automata, Phase Fields and Other Coarse-Grained Models

3.5.1 Cellular Automata

Individual cell models try to model as accurately as possible the interactions between cells in a growing cancer cell population. In this way, they provide accurate results for a relatively small number of cells, although the increase in available computer power is pushing this limit higher and higher. To model large-scale features of tumor growth, one can resort to different classes of models treating the kinetics at a coarse-grain scale, at the expense of the detailed modeling of the single cell. This can be done by using cellular automata, discrete models where each cell is assigned to a lattice site (Sottoriva et al., 2010; Jiao and Torquato, 2011, 2012, 2013; Chen et al., 2014b; La Porta et al., 2015a; Waclaw et al., 2015).

Probably the simplest possible cellular automaton considers only cell duplication and diffusion in a square lattice in two dimensions (La Porta et al., 2015a). Each site of a square lattice can be either occupied by a cancer cell or empty. At each step, one cell is picked at random and, if at least one nearest-neighbor site is empty, it can generate another cell with rate k_{diff} or diffuse with rate k_{div} (Figure 3.11a). Depending on the ratio k_{diff}/k_{div}, the model would predict different growth patterns for cancer cell colonies. In particular, if diffusion is faster than division we observe a fuzzy front (Figure 3.11b), otherwise fronts are less fuzzy and the colony grows faster (Figure 3.11c).

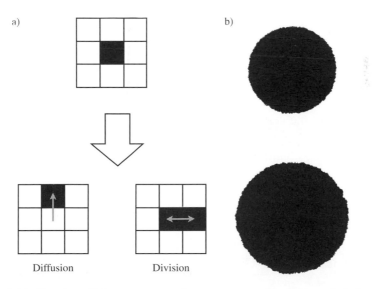

Figure 3.11 Simple cellular automaton for cancer growth. a) Occupied sites in a square lattice can either diffuse or divide. b) An example of the growth of a colony from the cellular automaton. Image by Lasse Laurson.

More elaborate cellular automata have been studied in the literature with the aim to model with increasing realism tumor growth (Sottoriva et al., 2010; Jiao and Torquato, 2011, 2012, 2013; Chen et al., 2014b; Waclaw et al., 2015): The simple square lattice can be replaced by random lattices in two and three dimensions and one can introduce different cellular types (Sottoriva et al., 2010) or more complex rules of evolution (Waclaw et al., 2015). The advantage of cellular automata lies in the possibility to run large scale simulations for long times with relatively little computational effort.

3.5.2 Continuous Models for Tumor Growth

Another class of coarse-grained models that are widely used to simulate cancer growth is based on multiple phase-fields (Lowengrub et al., 2010; Ben Amar et al., 2011; Wu et al., 2013; Chatelain et al., 2011). In these models, one writes reaction diffusion equations for the concentrations of cancer cells, the ECM, nutrients such as glucose or oxygen, and growth factors and inhibitors (for an extensive review see Lowengrub et al., 2010). The complexity of the model can be increased at will by taking care of more and more phases whose spatio-temporal evolution is described by phenomenological equations.

As an illustration, we describe here the model of Chatelain et al. (2011) describing a growing cellular phase, an interstitial liquid phase, and nutrient field with concentrations ϕ_c, ϕ_l and n, respectively.

The cellular and liquid phases obey convection-diffusion equations

$$\frac{\partial \phi_c}{\partial t} + \nabla \cdot (\mathbf{v}_c \phi_c) = \Gamma_c \tag{3.39}$$

$$\frac{\partial \phi_l}{\partial t} + \nabla \cdot (\mathbf{v}_l \phi_l) = \Gamma_l \tag{3.40}$$

where $\mathbf{v}_c$ and $\mathbf{v}_l$ are the velocity of the cellular and liquid phases and $\Gamma_c = -\Gamma_l$ are reaction rates. The authors impose a saturation condition $\phi_c + \phi_l = 1$, leading to an incompressibility condition $\nabla (\mathbf{v}_c \phi_c + \mathbf{v}_l \phi_l) = 1$.

The velocity of the cellular phase is then chosen to obey the Darcy law

$$\mathbf{v}_c = -D(1 - \phi^2)\nabla\Sigma, \tag{3.41}$$

with concentration-dependent porosity. The stress Σ is expressed in terms of phenomenological Landau–Ginzburg-type free energy functional of the field ϕ:

$$\Sigma = f(\phi_c) - \epsilon \nabla^2 \phi_c. \tag{3.42}$$

Finally, the reaction rate couples the growth of the cellular phase to the nutrient concentration

$$\Gamma_c = \gamma_c \phi_c (n/n_s - \delta_c), \tag{3.43}$$

where n_s is the typical nutrient concentration inside the tissue.

Nutrients are described by a simple diffusion-consumption equation (Ben Amar et al., 2011), which is assumed to be very fast, so that it can be described by an equilibrium condition

$$D_n \nabla^2 n - \delta_n \phi_c n + S_n (n_s - n) = 0, \tag{3.44}$$

where D_n is the diffusion constant, δ_n is a reaction parameter and S_n is the diffusion constant from the source of nutrients. Using this model it is possible to show that growing tumors display contour instabilities (Ben Amar et al., 2011) and micro-structural patterns (Chatelain et al., 2011) that are reminiscent of those observed skin tumors.

4

Vascular Hydrodynamics and Tumor Angiogenesis

Tumors need nutrients and oxygen to grow and, at the same time, must evacuate metabolic waste and carbon dioxide. To this end, tumor cells induce the sprout of new vessels from the pre-existing ones through a process known as angiogenesis, introduced in Section 4.1, where we mostly focus on the main biological aspects. In Section 4.2, we discuss how tumor cells can form vessels directly, a process known as *vasculogenic mimicry*. The physical aspects of angiogenesis, such as the constraints posed by the hydrodynamics of blood flow, are reviewed in Section 4.3. Finally, Section 4.4 is devoted to a discussion of the physics-based computational models for angiogenesis. These models are interesting since they allow therapeutic strategies in silico to be tested by simulating the release of drugs through the blood flow.

4.1 Biological Aspects of Angiogenesis

The growth of a tumor and its capacity to invade and give rise to metastasis crucially depend on angiogenesis, which is thus considered one of the hallmarks of cancer (Hanahan and Weinberg, 2000). Whereas blood vessel growth is tightly regulated under physiological conditions, the vasculature produced by angiogenesis in tumors typically displays aberrant structure, geometry and organization (see Figure 4.1), showing an excess of branching and modifications in the shape and size of vessels, which is reflected in defective blood flow and leakage (McDonald and Choyke, 2003; Nagy et al., 2010).

The acquisition of an angiogenic phenotype in tumors is associated to a switch in the balance between pro- and anti-angiogenic factors (Baeriswyl and Christofori, 2009; Bergers and Benjamin, 2003). This switch can be specific to distinct tumor types and localizations, and might also be altered during tumor progression (Folkman, 2002). Well-known factors inducing and inhibiting angiogenesis are the vascular endothelial growth factor-A (VEGF-A) and thrombospondin-1

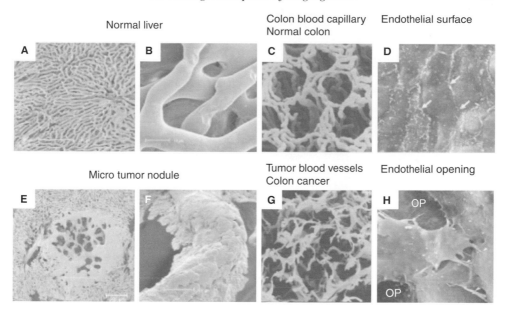

Figure 4.1 Comparison of SEM images of blood vessels in normal healthy tissues (AD) and in tumor tissues (EH) of metastatic tumor nodules in the liver from colon cancer. Image by Maeda, 2012, CC-BY-2.0 licence.

(TSP-1), respectively (Figure 4.2). The VEGF-A gene encodes ligands involved in the generation and organization of blood vessels during development and also during adult life, when they are needed for the survival of endothelial cells. The regulation of VEGF signaling is very complex, reflecting the variety of its functions. For instance, the level of VEGF can be upregulated by the lack of oxygen (hypoxia) or by oncogene signalling (Ferrara, 2009; Mac Gabhann and Popel, 2008; Carmeliet, 2005). VEGF-A ligands can be found in the ECM in latent forms that can be activated by ECM-degrading proteases (e.g., MMP-9, Kessenbrock et al., 2010). Other pro-angiogenic signals also exist, such as, for instance, members of the fibroblast growth factor (FGF) family, which can sustain angiogenesis when they are chronically overexpressed (Baeriswyl and Christofori, 2009). Angiogenesis is inhibited by TSP-1, which binds to endothelial cells transmembrane receptors that send signals opposing the action of VEGF (Kessenbrock et al., 2010). In addition to VEGF, a network of interconnected signaling pathways involving ligands of signal-transducing receptors displayed by endothelial cells (e.g., Notch, Neuropilin, Robo, and Eph-A/B) has recently been shown to be deeply involved in angiogenesis. These pathways have been functionally implicated in developmental and tumor-associated angiogenesis and illustrate the complex regulation of endothelial cell phenotypes (Carmeliet and Jain, 2000; Dejana et al., 2009; Pasquale, 2010). Interestingly, distinctive gene expression profiles were shown

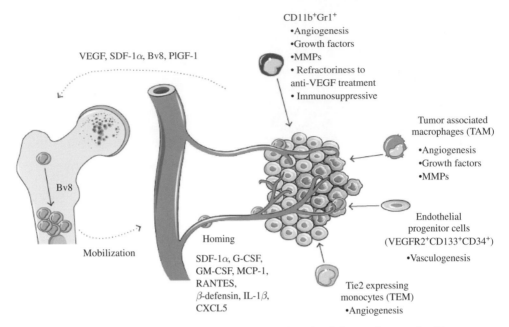

Figure 4.2 Schematic of the main factors involved in angiogenesis. Tumor and stromal cells mobilize various subpopulations of tumor, promoting bone marrow-derived cells to the peripheral blood through secretion of cytokines and chemokines. Diverse chemoattractant factors promote the recruitment and infiltration of these cells to the tumor microenvironment where they promote tumor angiogenesis and vasculogenesis. Image by Schmid and Varner, 2010, CC-BY-3.0 licence.

to be tumor-associated to endothelial cells (Nagy et al., 2010; Ruoslahti et al., 2010).

In addition to blood vessels, the formation of lymphatic vessels (lymphangiogenesis) has also been investigated extensively in recent years (Tammela and Alitalo, 2010). Vascular remodelling associated with lymphangiogenesis and angiogenesis actually involve similar processes. In response to molecular mediators, both lymphatic and vascular endothelial cells first proliferate and migrate towards the tumor while the ECM is degraded and then endothelial cells arrange into tube-like structures (Sleeman et al., 2001). Furthermore, new production and realignment of the ECM and controlled apoptosis at appropriate sites are required for blood vascular and lymphatic system formation. Beside using similar processes of remodelling, blood and lymphatic vessels are closely associated in vivo. Progress in understanding lymphangiogenesis has been hampered by the close similarity of blood and lymphatic vessels in tissues and is confounded by the lack of lymphatic-specific markers (Kaipainen et al., 1995). More accurate and simplified lymphatic vessel identification has recently been made possible by the discovery of molecules that are specifically expressed by lymphatic endothelium, such as the vascular endothelial growth factor receptor-3 (VEGFR-3), predominantly expressed on the lymphatic endothelium in normal adult tissues (Kukk et al., 1996). In particular, VEGF-C and VEGF-D, two members of the VEGF family of secreted glycoproteins whose receptor is VEGFR-3 (Joukov et al., 1996), have been identified as regulators of lymphangiogenesis in mammals (Oh et al., 1997).

The lymphatic receptor for hyaluronan, LYVE-1, has been reported to be a specific marker of lymphatic vessels and is thought to function in transporting hyaluronan from the tissue to the lymph (Banerji et al., 1999; Prevo et al., 2001). The transcription factor Prox 1, although required for lymphatic vessel development and expressed in lymphatic endothelium (Wigle and Oliver, 1999), is also expressed in other cell types and tissues, including hepatocytes and liver (Sosa-Pineda et al., 2000) and lens tissue (Wigle et al., 1999), and is therefore of limited use immunohistochemically to identify lymphatic vessels. Podoplanin and desmoplakin have been reported to be markers for lymphatic endothelium but also react with other cell types (Weninger et al., 1999; Wells et al., 1997; Ebata et al., 2001). In summary, a more extensive range of markers for lymphatic endothelium is now available that should aid in defining the role of lymphatic vessels in tumor biology. However, their role in the tumor-associated stroma, specifically in supporting tumor growth, is poorly understood. Indeed, because of high interstitial pressure within solid tumors, intratumoral lymphatic vessels are typically collapsed and nonfunctional; in contrast, however, there are often functional, actively growing (lymphangiogenic) lymphatic vessels at the peripheries of tumors and in the adjacent normal tissues invaded by cancer cells. These associated lymphatics likely serve as channels for the seeding of metastases in the draining lymph nodes that are commonly observed in a number of cancer types. Several drugs inhibiting angiogenesis have been developed in the past few years, targeting, for instance, VEGF.

4.2 Vasculogenic Mimicry in Cancer

In addition to angiogenesis, recent data suggest that the development of tumour vasculature occurs through vasculogenesis too. Vasculogenesis is a spontaneous development of vessels through mobilisation, recruitment, differentiation and vascular incorporation of bone marrow (BM)-derived endothelial progenitor cells (EPCs) (Ding et al., 2008). Cancer cells have been shown to express an endothelial phenotype, forming vessel-like networks in three-dimensional cultures, mimicking the pattern of embryonic vascular networks (La Porta, 2011). This process is defined as "vasculogenic mimicry," where the word "vasculogenic" indicates the generation of a vascular network *de novo* and mimicry is used because the tumor uses the network for transporting fluids in tissues that are clearly not blood vessels. Vasculogenic mimicry was reported in melanoma and in several other tumors, including breast, prostate, ovarian, chorio-, lung carcinomas, synovial-rhabdomyosarcoma, Ewing sarcomas and paeochromocytoma (La Porta, 2011). Tumor cells capable of vasculogenic mimicry exhibit a high degree of plasticity indicative of a multipotent phenotype similar in many respects to embryonic stem cells (Bittner et al., 2000; Bissell, 1999; Hendrix et al., 2003).

Molecular profiling of the tumor cell vasculogenic mimicry phenotype has revealed highly upregulated genes associated with embryonic progenitors, endothelial cells, vessel formation, matrix remodeling, and hypoxia; and downregulated genes generally associated with the respective lineage-specific phenotype. It is known that tumor cells express key pluripotent stem cell markers. However, unlike normal embryonic progenitors, tumor cells lack major regulatory checkpoints, resulting in the aberrant activation of embryonic signaling pathways such as Nodal and Notch, giving rise to unregulated growth and aggressive behavior (Hardy et al., 2010). For instance, in the case of melanoma, several melanocyte-lineage genes are suppressed, as confirmed by comparing laser capture microdissection and microgenomics profiling of living melanoma cells with endothelial cells forming vascular networks (Hendrix et al., 2003). In particular, the expression of specific angiogenesis-related genes in melanoma resembles that of normal endothelial cells (Demou and Hendrix, 2008).

The presence of vasculogenic mimicry in tumor tissues from patients has been associated with a poor clinical outcome and suggests a possible advantage imparted by vasculogenic mimicry for the survival of aggressive tumor cell phenotypes. Experimental evidence shows indeed a physiological perfusion of blood between endothelial-lined mouse vasculature and vasculogenic mimicry networks in human tumor xenografts (Ruf et al., 2003). Additional studies identified the anticoagulant properties of tumor cells which line vasculogenic mimicry networks. Thus, vasculogenic mimicry can provide a functional perfusion pathway for rapidly growing tumors by transporting fluid from leaky vessels and/or connecting with endothelial-lined vasculature. A remarkable example of the functional plasticity provided by vasculogenic mimicry was achieved by transplanting human metastatic melanoma cells into a circulation-deficient mouse limb, resulting in the formation of a human melanoma-mouse endothelial chimeric neovasculature (Hendrix et al., 2002). After the restoration of blood flow to the limb, tumor cells formed a large tumor mass. Hence, this study highlights the powerful influence of the microenvironment on the transendothelial differentiation of melanoma cells, which revert to a tumorigenic phenotype in response to changed environmental cues.

4.3 Physical Aspects of Vascular Flow in Tumors

Most of the enormous literature devoted to tumor angiogenesis and vasculogenic mimicry focuses on the identification of key factors regulating these processes. The physical aspects of vessel transformation and remodeling are, however, equally important, but much less studied and understood. The formation and modification of a blood vessel network by growing tumors is a complex process involving collective mechanical and hydrodynamical forces. Vessels react to mechanical stimuli due to the flowing blood, namely radial pressure and shear stresses (Figure 4.3). The reaction often involves a feedback controlling blood flow or adapting the vessel structure to the required function.

In a first approximation, blood flow on a vessel of radius r and length l obeys the ideal Hagen–Poiseuilles law stating that the flow rate q is given by

$$q = \frac{\pi r^4 \Delta p}{8 \eta l},\tag{4.1}$$

where η is the blood viscosity and Δp is the pressure difference between the two ends of the vessel. The fourth power dependence implies that even small variations in the vessel radius can yield very large changes in the flow. Furthermore, the shear stress in the vessel wall is given by

$$\sigma = \frac{r \Delta p}{2l}.\tag{4.2}$$

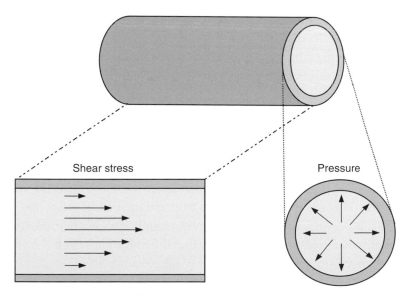

Figure 4.3 Mechanical stresses due to flow in a vessel can be decomposed into shear stresses and radial pressures.

We have already mentioned that the vascular structure observed in tumors is considerably different from that in normal tissues (see Figure 4.1). The normal blood vessels form a well-organized architecture consisting of arterioles, capillaries and venules (Fukumura and Jain, 2007). In contrast, tumor vessels are dilated, tortuous and disorganized in their patterns of interconnection (Jain, 1988). Normal vasculature is characterized by dichotomous branching, but tumor vasculature is unorganized and can present trifurcations and branches with uneven diameters. Perivascular cells in tumor vessels have abnormal morphology and heterogeneous association with vessels (McDonald and Choyke, 2003). In short, the hierarchical organization of the normal vasculature into an arterio-venous architecture is completely missing in tumor vasculature (Welter and Rieger, 2010). As a result of this, the resistance to blood flow in tumors is increased, leading to lower perfusion rates (blood flow rate per unit volume) in tumors with respect to normal tissues. Tumor blood flow is also unevenly distributed, fluctuates with time and can even reverse its direction in some vessels.

The fluid present in tissues outside of vessels is usually referred to as interstitial. Unlike normal tissues, in which the interstitial fluid pressure (IFP) is around zero, both animal and human tumors exhibit interstitial hypertension (Jain and Duda, 2004). Two major mechanisms contribute to interstitial hypertension in tumors. In normal tissues, the lymphatics maintain fluid homeostasis; thus, the lack of functional lymphatics in tumors is a key contributor, as drainage of excess fluid from the tumor interstitium is impaired. The net result is accumulation of fluid and, hence, increased fluid pressure in the interstitial space. Indeed, DiResta et al. (2000) were able to lower the IFP by placing artificial lymphatics in tumors. The second contributor is the high permeability of tumor vessels. The tumor IFP begins to increase as soon as the host vessels become leaky: this increases the hydrostatic pressure in the tumor vessels and drives the flow of fluid from the vascular to the interstitial space. Thus, the interstitial fluid may ooze out of the tumor into the surrounding normal tissue, carrying away chemotherapeutic drugs with it (Netti et al., 2000; Heldin et al., 2004; Baxter and Jain, 1989; Salnikov et al., 2003). These findings have been confirmed also by computational continuum models (Wu et al., 2013; Welter and Rieger, 2013). The reverse situation can also arise: when the local IFP overcomes the pressure inside the vessel, blood flows from the interstitial space into the leaky vessels. This, however, can only occur locally, as an IFP higher than the hydrostatic intravascular blood pressure is not sustainable. IFP indeed has been shown to go up and down with the microvascular pressure within seconds (Netti et al., 1995). Furthermore, IFP has been found to be nearly uniform throughout tumors, dropping precipitously in the tumor margin (Rofstad et al., 2002; Boucher et al., 1990).

Finally, collapsed lymphatic and blood vessels contribute to elevated IFP in tumors and not the other way around (Boucher and Jain, 1992; Griffon-Etienne et al., 1999; Padera et al., 2004; Helmlinger et al., 1997). Therefore, solid stress cannot be affected by IFP (Stylianopoulos et al., 2012); nor is the opposite true. However, they can, by contrast, cause compression of fragile blood vessels, resulting in poor perfusion and hypoxia (Stylianopoulos et al., 2012, 2013). Elevated compressive solid stress in the interior of tumors can be sufficient to cause blood vessel collapse, resulting in a lower growth rate of cancer cells with respect to the periphery. Notice, however, that vessels can also collapse due to other mechanisms (Holash et al., 1999).

4.4 Computational Models of Vascular Remodeling

Physics-based computational models of tumor angiogenesis are playing an important role in the improved understanding of these phenomena. Detailed and realistic models provide an innovative strategy for testing therapeutic interventions *in silico*, such as the delivery of drugs to cancer cells through the blood flow, and could help develop diagnostic tools based on the simulation of the distribution of oxygen and nutrients within the tumor (see Rieger and Welter, 2015, for a recent review).

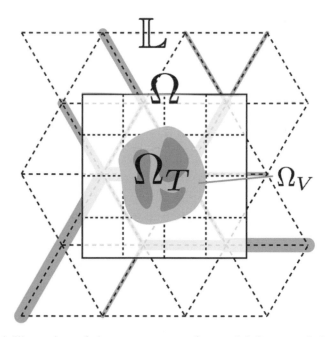

Figure 4.4 Illustration of the components of a model for vascularized tumor growth. L denotes the lattice whose edges can be occupied with vessel segments. Ω denotes the region where continuum equations are defined. The tumor region Ω_T is indicated in grey. The darker tone indicates necrotic regions. Viable regions, denoted Ω_V, are brighter. Image by Welter and Rieger, 2013, CC-BY-4.0 license.

(a)   (b)

(c)   (d)

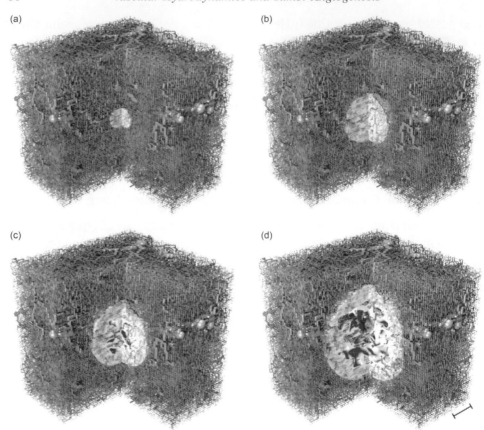

Figure 4.5 Snapshots from the simulation of the growth of a vascularized tumor
according to the model of Welter and Rieger (2013). Image by H. Rieger, 2013.

Building a realistic computational model of a growing vascularized solid tumor
is a complex undertaking requiring several steps (Rieger and Welter, 2015). It is
useful to divide a model into different components: (i) a model for tumor growth,
(ii) a model for vascular remodeling, (iii) a model for oxygen transport, extravasation and diffusion, (iv) a model for fluid extravasation and interstitial flow and
(v) a model for drug transport and release. While several computational models of
vascularized tumor growth have been proposed in the literature (Araujo and McElwain, 2004), complete models involving all the components described above have
appeared only recently (Welter and Rieger, 2010, 2013).

An outline of the model decomposition is reported in Figure 4.4, showing a
tumor region, divided into a necrotic core and a viable part, surrounded by an
active region where one defines transport equations and finally a set of discretized
vascular channels (Welter and Rieger, 2013). To build the tumor growth component of the model one can follow the strategies outlined in Sections 3.4 and 3.5,

using either individual cell models or more coarse-grained approaches provided by cellular automata or phase-field model. Simulation of large tumors inevitably requires the use of coarse-grained models (Welter and Rieger, 2010, 2013). The growth rules of tumor cells should be coupled to the concentration of oxygen and nutrients, which in turn is dictated by the solution of equations for transport, diffusion and extravasation. Those equations also depend on the vessel structure that is constantly remodeling as the tumor grows.

Vascular remodeling can be simulated by initially placing the vascular network on a regular grid in two or three dimensions (Bartha and Rieger, 2006; Lee et al., 2006). Flow in the network can be computed by numerically solving the Kirchhoff equation, imposing that the total flow at each node is zero (i.e., $\sum_j q_{i,j} = 0$, where $q_{i,j}$ is the flow between nodes i and j and the sum is restricted to all nodes j connected to node i), with appropriate boundary conditions reflecting experimental data. To avoid spurious artefacts due to an artificial initial vessel geometry, Welter and Rieger (2013) generate realistic vessel network by imposing physiological statistical constraints consistent with real vasculatures. As the tumor grows, the initial networks change due to sprout initiation and migration, vessel wall degeneration, vessel collapse and dilation. All these processes are implemented in discrete time by specific rules depending on the blood flow rate, the shear stress and the interstitial fluid flow that is also explicitly treated by solving equations describing a liquid in a porous medium (Welter and Rieger, 2013).

An illustration of the results obtained with an integrated complex model of vascularized tumor growth is given in Figure 4.5, showing the combined time evolution of the tumor and the vasculature. Numerical simulations of these models provide useful information to assess the validity of therapeutic strategies. For instance, Rieger and Welter (2015) show that a theory targeting only vessel leakiness is unlikely to be effective, while increasing the permeability of tumor tissues can improve drug delivery.

5

Cancer Stem Cells and the Population Dynamics of Tumors

According to the CSC hypothesis, tumor growth is driven by a small population of cancer stem cells which can self-renew and differentiate into cancer cells with a long but limited lifespan. The experimental challenges in the identification of CSCs are discussed in Section 5.1 where we review the relevant biological markers. In Section 5.2, we discuss the population dynamics within the tumor, composed by CSCs and other tumor cells. To this end, we introduce a mathematical model for the evolution of the populations and compare its results with experiments. In Section 5.3, we deal with the possibility that cancer cells can occasionally switch back to the CSC state, discussing relevant experiments and models. Section 5.4 is devoted to the issue of cell sorting. While one would like to define the CSC population in a sharp way, in practice the boundary is fuzzy, being based on the expression of biological markers which are by their nature imperfect. This opens an interesting issue about the interpretation of the experimental results.

5.1 Experimental Identification of Cancer Stem Cells

Tumor growth can either be described by the conventional model or by the CSC theory. According to the first model, cancer cells are heterogeneous but are all tumorigenic, while the CSC hypothesis states that in the tumor there is a subpopulation sustaining the tumor growth (La Porta, 2009). Cancer stem cells (CSCs) are a subpopulation of tumor cells that possess the stem cell properties of self-renewal and differentiation. Therefore, a CSC is a cell within the tumor that has the capacity of self-renewing and generating a heterogeneous population of cancer cells composing the tumor. Contrary to normal stem cells, however, which control very strictly their proliferation capability and try to carefully maintain their genomic integrity, CSCs typically lack any control of those processes. Identifying differences between normal stem cells and CSCs is important for understanding how cancers progress and could possibly result in new therapies to fight cancer.

In practice, CSCs can only be defined experimentally by their ability to recapitu-late the generation of a continuously growing tumor. To this end, putative CSCs are identified according to the expression of surface markers (e.g., factors expressed by normal stem cells) and isolated through fluorescence-activated cell sorting (FACS). The cells isolated in this way are then transplanted (or engrafted) in immunocom-promised mice. When murine cancer cells are transplanted into a mouse, we talk about syngenic transplantation, while the transplant of human cancer cells into a mouse is called a xenograft. If the mouse develops a tumor, cancer cells expressing the CSC markers are again isolated and transplanted into mice (Figure 5.1). The observation of a series of CSC identifications and transplantations is considered evidence for the existence of a CSC population within the tumor.

Several aspects related to the identification of CSCs are, however, problematic. The first problem stems from the use of serial transplantation to validate a candidate CSC subpopulation by monitoring the capability to recapitulate the heterogeneity of the primary tumor. Both xeno- and syngeneic transplantation might misrepre-sent the intricate network of interactions with diverse support such as fibroblasts, endothelial cells, macrophages, mesenchymal stem cells and many of the cytokines

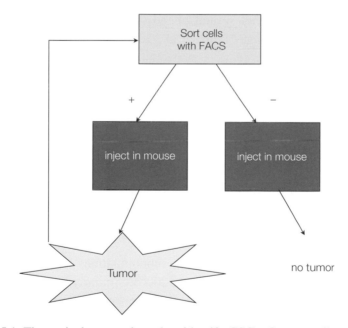

Figure 5.1 The typical protocol used to identify CSCs. Cancer cells are sorted in positive (+) and negative (−) populations according to the expression of some marker. Positive and negative cells are then transplanted in mice. If the marker selects CSCs, then a tumor should result only for positive cells. If this is the case, positive cells are identified again in the tumor and the process is repeated.

and receptors involved in these interactions (for a more comprehensive discussion, read La Porta, 2009). Another important problem is that there is no clear recipe to choose the best marker to unambiguously identify CSCs. One is thus forced to proceed by trial and error, exploiting analogies with normal stem cells. The general aim is to use these markers to develop therapeutic strategies to target CSCs. For instance, the Notch receptor signaling pathway has been extensively studied as a possible target to hit CSCs. This pathway is, however, active in many tissues and therefore a possible treatment might display a great toxicity. It is therefore imperative to first understand better the complex regulation associated with this pathway (Harrison et al., 2010).

The first evidence of CSCs came from hematological tumors (Bonnet and Dick, 1997) and later from solid tumors such as breast, prostate, brain cancer and melanoma. In breast cancer, the first evidence of a subpopulation with a specific cell-surface antigen profile (CD44+/CD24−) that successfully established itself as tumor xenograft was published in 2003 (Al-Hajj et al., 2003). In another study, 275 samples obtained from patients with primary breast cancers of different subtypes and histological stages were analyzed for CD44+CD24− putative stem cell markers as well as for other markers (vimentin, ostenectin, connexin 43, ADLH, CK18, GATA3, MUC1). The study revealed a high degree of diversity in the expression of several of the selected markers in different tumor subtypes and histologic stages (Park et al., 2010). Furthermore, aldehyde dehydrogenase (ALDH) was used as a stem cell marker in 33 human breast cell lines (Charafe-Jauffret et al., 2009). ALDH is a detoxifying enzyme that oxidizes intracellular aldehydes and is thought to play a role in the differentiation of stem cells via the metabolism of retinal to retinoic acid (Chute et al., 2006). Interestingly, ALDH activity can be used to sort a subpopulation of cells that display stem cell properties from normal breast tissue and breast cancer (Ginestier et al., 2007) and to isolate CSCs from multiple myeloma and acute leukemia as well as from brain tumors (Cheung et al., 2007; Corti et al., 2006). The ALDH phenotype was not associated with more-aggressive subpopulations in melanoma, suggesting that it is not a "universal" marker (Prasmickaite et al., 2010).

Several markers that select aggressive subpopulations have been identified in glioblastoma multiforme (Salmaggi et al., 2006). Similarly, several candidate populations of prostate stem/progenitor cells have been reported, including those expressing high levels of CD44, integrin $\alpha2\beta1$, or CD133 (Tang et al., 2007). Interestingly, two recent independent studies in the mouse prostate have identified two different populations of stem cells. One, marked by CD117 (c-Kit), seems to be localized in the basal layer (Leong et al., 2008) and the other, called castration-resistant Nkx3.1-expressing cells, in the luminal layer (Wang et al.,

2009). Identification and characterization of normal prostate stem cells is clearly relevant to understanding the origin for human prostatic cancer (Li and Tang, 2011). This is because it is difficult to ascertain the potential overlap as well as lineage relationships of the various candidate stem cells that have been identified (Shen and Abate-Shen, 2010). This in part is due to the distinct methodologies and assays employed (Shen and Abate-Shen, 2010).

Several papers came out between 2005 and 2008 reporting evidence for the existence of a CSC subpopulation in melanoma (Fang et al., 2005; Hadnagy et al., 2006; Dou et al., 2007; Monzani et al., 2007; Schatton et al., 2008; Klein et al., 2007). In 2008, however, a paper argued against CSCs based on the following observations: a relatively large fraction of melanoma cells (up to 25 percent) was shown to initiate tumors in severely immunocompromised NOD/SCID IL2Rγnull mice; the fraction of tumor-inducing cells depended upon assay conditions; several putative CSC markers appear to be reversibly expressed (Quintana et al., 2008). This paper suggests that the best experimental model to confirm the presence of CSCs is severely immunocompromised mice. The authors analyzed the expression of more than 50 surface markers on melanoma cells derived from several patients (A2B5, cKIT, CD44, CD49B, CD49D, CD49F, CD133, CD166) but then mostly focused on CD133 and CD166 (Quintana et al., 2008). However, in a more recent paper it was shown that CD133 is highly expressed in melanoma cells, and it is not a good marker to sort CSCs (Schatton et al., 2008). Moreover, in 2010, an independent group, using the same immunocompromised mice, did not confirm the results of Quintana et al. (2010), and proposed instead to use CD271, the nerve growth factor receptor, as a marker to identify CSCs (Boiko et al., 2010). Finally, CXCR6 was proposed as a marker for a CSC-like aggressive subpopulation in human melanoma cells (Taghizadeh et al., 2010). Its peculiar feature is that it plays a critical role in stem cell biology, being linked to asymmetric cell division (Taghizadeh et al., 2010).

Recent experiments in vivo have confirmed the presence of an aggressive CSC-like subpopulation in benign and malignant intestinal and skin tumors (Schepers et al., 2012; Driessens et al., 2012; Chen et al., 2012). Contrary to previous work based on sorting and serial transplantation, these new papers tracked CSCs directly inside a growing tumor and studied the CSC population at different stages of tumor progression.

5.2 Population Dynamics of Cancer Stem Cells

According to the CSC hypothesis, cells are organized hierarchically, with CSCs at the top of the tree. The population dynamics of cancer cells in the CSC model can be naturally described by the theory of branching processes (see Section 3.1) (Michor et al., 2005; Ashkenazi et al., 2008; Tomasetti and Levy, 2010; La Porta et al., 2012; Zapperi and La Porta, 2012).

As an illustration of this approach we follow La Porta et al. (2012), considering a model in which CSCs can divide symmetrically with rate R_d giving rise to two new CSCs with probability p_2, two cancer cells (CCs) with probability p_0, or divide asymmetrically with probability $p_1 = 1 - p_0 - p_2$ giving rise to a CSC and a CC. CSCs can duplicate for an indefinite amount of time, CCs become senescent after a finite number of generations M and then eventually die with rate q (Figure 5.2). The kinetics of the model depends only on the combination $\epsilon = p_2 - p_0$, the average relative increase in the number of CSCs after one duplication, and not on the individual parameters p_0, p_1 and p_2. In normal tissues, the stem cell population should remain constant, which implies that $\epsilon = 0$ (Ashkenazi et al., 2008; Clayton et al., 2007), while in tumors we expect $\epsilon > 0$.

The kinetics of the cell populations for the model can be solved exactly. To this end, we first derive recursion relations linking the average cell populations at each generation. Denoting by S^N the average number of CSCs after N generations, by C_k^N the average number of CCs, where $k = 1, \ldots, M$ indicates the "age" of the CCs (i.e., the number of generations separating it from the CSC from which it originated), and by D^N the average number of senescent cells, we obtain

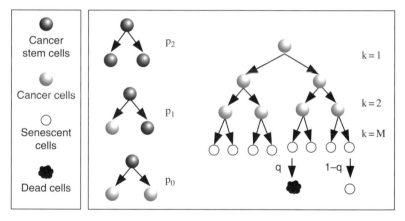

Figure 5.2 Branching process for cancer stem cells: at each generation, CSCs can divide symmetrically, giving rise to two CSCs with probability p_2 or to two CCs with probability p_0, or asymmetrically with probability $p_1 = 1 - p_2 - p_0$ giving rise to a CSC and a CC. Cancer cells can only divide a finite number of times M (in the figure $M = 3$); after that they become senescent. Senescent cells die with probability q at each generation. Image by La Porta et al., 2012, CC-BY-3.0 licence.

$$S^N = (1 + \epsilon)S^{N-1}$$
$$C_1^N = (1 - \epsilon)S^{N-1} \tag{5.1}$$

$$\cdots \cdots \cdots$$

$$C_k^N = 2C_{k-1}^{N-1}$$
$$D^N = (1 - q)D^{N-1} + 2C_M^{N-1}.$$

These relations are derived considering that CSCs can only originate from other CSCs either by a symmetric division – two CSCs are generated with probability p_2 – or by an asymmetric one, in which case a single CSC is generated with probability $1 - p_2 - p_0$. Hence each CSC generates an average of $2p_2 + 1 - p_2 - p_0 = 1 + \epsilon$ new CSCs. CSCs also generate CCs ($k = 1$) by asymmetric CSC divisions with probability $1 - p_2 - p_0$ or by symmetric CSC division with probability p_0, yielding an average of $1 - p_2 - p_0 + 2p_0 = 1 - \epsilon$ CCs. Two CCs are generated by duplication of other normal cells with unit probability ($k = 2 \ldots M$). Senescent cells accumulate at each generation when normal cells (with $k = M$) lose the ability to duplicate and die with probability q.

Equations 5.2 can be solved explicitly. Consider first an initial condition with CSC only (i.e., $C_k^0 = D^0 = 0$), yielding

$$S^N = (1 + \epsilon)^N S^0$$

$$C_k^N = \begin{cases} S^N \left(\frac{2}{1+\epsilon}\right)^{k-1} \frac{1-\epsilon}{1+\epsilon} & \text{for } N > k \\ 0 & \text{for } N \leq k \end{cases}$$

$$D^N = \begin{cases} S^N \frac{(1-\epsilon)}{\epsilon+q} \left(\frac{2}{1+\epsilon}\right)^M \left(1 - \left(\frac{1-q}{1+\epsilon}\right)^{N-M}\right) & \text{for } N > M \\ 0 & \text{for } N \leq M. \end{cases} \tag{5.2}$$

Equations 5.2 imply that the average CSC population S^N grows exponentially fast, and that after a large number of generations (i.e., for $N \gg M$), all cell populations, and hence the size of the tumor, are proportional to the CSC population. This result confirms that the population of CSCs is the driving force behind tumor growth. Note that the population of senescent cells is also driven by the growth of CSCs, and therefore cancer growth and senescence are inextricably linked.

We can also consider the case in which the initial condition is composed of a mixed population of CSCs and CCs. The solution in this case can be obtained as a linear superposition of the solution of Equations 5.2 and the solution with initial conditions $S^0 = 0$, $C_k^0 > 0$ and $D^0 > 0$. For the case of a uniform distribution for the ages of CSCs: $C_k^0 = D^0 = c$, we obtain

$$\hat{S}^N = 0$$

$$\hat{C}_k^N = \begin{cases} 2^N c & \text{for } N < k \\ 0 & \text{for } N \geq k \end{cases}$$

$$\hat{D}^N = \begin{cases} c(1-q)^N + c\frac{2}{1+q}(2^N - (1-q)^N) & \text{for } N \leq M \\ c(1-q)^N (1 + \frac{2}{1+q}((\frac{2}{1-q})^M - 1)) & \text{for } N > M. \end{cases} \qquad (5.3)$$

The complete solution for the total number of cells, corresponding to an initial condition $S^0 > 0$ and $C_k^0 = D^0 = c$, is given by

$$n_{tot}^N = S^N + \hat{S}^N + \sum_{k=1}^{M}(C_k^N + \hat{C}_k^N) + D^N + \hat{D}^N, \qquad (5.4)$$

and is plotted in Figure 5.3a, while in Figure 5.3b we report the fraction of senescent cells. These theoretical predictions closely match what is observed in experiments when melanoma cells are first sorted according to the ABCG2 marker (Monzani et al., 2007; Taghizadeh et al., 2010) and then grown for several months in vitro (La Porta et al., 2012). ABCG2+ cells grow substantially more than ABCG2− cells that almost stop growing after 100 days and then resume. At the same time, the fraction of senescence displays a peak in correspondance with the slowing down of the growth curve, as displayed by the model (5.3b). The natural interpretation is that ABCG2 selects a CSC rich subpopulation, while ABCG2− cells contain only a few CSCs.

Equations 5.2 represent only a transient contribution to the cell population and do not influence the asymptotic fractions (i.e., for $N \gg M$) of senescent cells f_{SC}^{∞} and CSCs f_{CSC}^{∞}, which are readily obtained by dividing the results in Equations 5.2 by the total number of cells at each generation:

$$f_{SC}^{\infty} = \frac{1-\epsilon}{1+q} \qquad (5.5)$$

$$f_{CSC}^{\infty} = \frac{q+\epsilon}{1+q}\left(\frac{1+\epsilon}{2}\right)^M. \qquad (5.6)$$

It is instructive to use the solution to compare how cells become senescent in normal tissues and in cancers. In the long time limit, the asymptotic fraction of senescent cells is equal to $f_S^{\infty} = (1-\epsilon)/(1+q)$ and is independent of the number of duplications M needed to induce senescence. For normal stem cells, $\epsilon = 0$ and the percentage of senescent cells is expected to be large, reaching 100 percent in the limit of $q = 0$, when senescent cells never die. For cancer cells, $\epsilon > 0$ and the fraction of senescent cells is smaller but still non-vanishing. Similarly, the CSC fraction is given by $f_{CSC}^{\infty} = \frac{q+\epsilon}{1+q}\left(\frac{1+\epsilon}{2}\right)^M$ which is smallest for normal stem cells and increases as a function of ϵ in tumors. Aggressive tumors should be characterized by high values of ϵ and therefore by relatively high values of the CSC fraction.

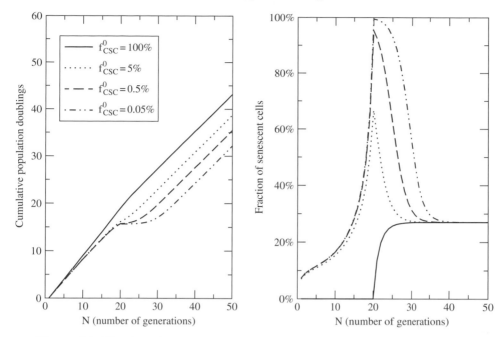

Figure 5.3 Kinetics of the CSC model. a) The growth of the cell population obtained in the model for $\epsilon = 0.7$, $q = 0.1$, $M = 20$ and different values of the initial fraction of CSCs f_{CSC}^0 as a function of the number of generations N. b) The evolution of the fraction of senescent cells corresponding to the same parameters. Image by La Porta et al., 2012, CC-BY-3.0 licence.

5.3 Phenotypic Switching

Recent results indicate that the population dynamics of CSCs is more complex than the strict hierarchy originally proposed. Experimental evidence indicate that non-CSCs breast cancer cells can revert to a stem-cell-like state even in the absence of mutations (Gupta et al., 2011). Similarly, in melanoma a small population of CSC-like JARID1B positive cells has been shown to be dynamically regulated in a way that differs from the standard hierarchical CSC model (Roesch et al., 2010), possibly reconciling earlier controversial results (Quintana et al., 2008, 2010; Boiko et al., 2010). Microenvironmental factors, such as TGFβ, are found to enhance the rate of switching from non-CSC cells to the CSC state (Chaffer et al., 2013).

This is in line with earlier results showing that ABCG2 negative cells isolated from human melanoma biopsies express this marker again after a few generations in vitro (Monzani et al., 2007). The idea that the environment is able to induce cells to switch into a more aggressive phenotype was explored by Medema and Vermeulen (2011), who were able to limit stemness by modifying microenvironmental factors known to support CSCs in tumors. While mounting experimental evidence supports the switch from non-CSC cancer cells to the CSC state, the biological factors regulating this process are only starting to be uncovered. Two possible scenarios have been invoked to solve this puzzle: the switch to the CSC state is either i) driven by genetic mutations or ii) regulated by epigenetic factors (Marjanovic et al., 2013).

We recently reported results indicating that the switching to the CSC state in human melanoma cells is regulated by an internal underdamped homeostatic mechanism controlling the fractional population, not dependent on external microenvironment (Sellerio et al., 2015). We measured the switching to the CSC state on the single-cell level with three biomarkers (CXCR6, ABCG2 and CD271) (Monzani et al., 2007; Taghizadeh et al., 2010; Boiko et al., 2010). For all three markers, positive cells are known to give rise to a bigger tumor in immunodeficient mice than the negative ones (Monzani et al., 2007; Taghizadeh et al., 2010; Boiko et al., 2010). We observed that reducing the percentage of CSCs below a threshold induces cells to switch en masse, substantially overshooting the fraction seen in unsorted cells. The regulatory mechanism underlying the observed massive phenotypic switching was traced back to the activity of a complex miRNAs network (see Section 1.6) which is differentially expressed as the CSC are switching (Sellerio et al., 2015), see Figure 5.4.

In order to understand how an overshoot in the expression of CSC markers can possibly arise after sorting, one can consider a mathematical model similar to the one introduced above. If we simply introduce a small probability for cancer cells to switch back to the CSC state (Gupta et al., 2011; Zapperi and La Porta, 2012), we would predict that the CSC population will re-approach its steady-state concentration after sorting, but without an overshoot. In order to observe an overshoot, the switch probability must depend on the expression level of a set of relevant miRNAs, which is in turn controlled by the population of CSCs. When the CSC population is depleted, the miRNA expression level shoots up, triggering a large switching probability (Sellerio et al., 2015). This induces an overshoot in the fraction of CSCs in the population that then falls back to the steady-state level as the level of miRNAs decreases.

In practice, one can consider a simple model for CSC population dynamics including CSCs S, duplicating cancer cells C and non-duplicating (senescent) cells D (Sellerio et al., 2015). The model kinetics assumes that each CSC can duplicate, yielding two CSC with probability p_2 or two cancer cells with probability p_0, or can divide asymmetrically into one CSC and one cancer cell with probability p_1, such that $p_0 + p_1 + p_2 = 1$. Cancer cells can switch to CSCs with a probability $p(\mu)$ depending on the concentration of relevant miRNAs μ, or duplicate and

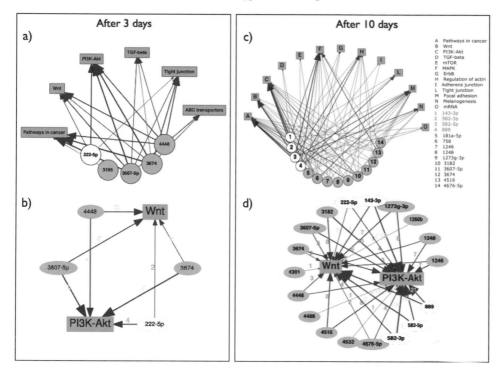

Figure 5.4 Interaction network between miRNAs and pathways during CSC switch in melanoma. a) The network formed by differentially expressed miRNAs in CXCR6-negative cells before the overshoot (3 days after sorting) and pathways identified to be significantly targeted by those miRNAs. b) The same network as in a) restricted to Wnt and PI3K signalling pathways. c) The network formed by differentially expressed miRNAs in CXCR6-negative cells at the overshoot (10 days after sorting) and pathways identified to be significantly targeted by those miRNAs d) The same network as in c) restricted to Wnt and PI3K signalling pathways. For all panels the miRNAs are shown in white when they increase and in grey when they decrease. From Sellerio et al. (2015) CC-BY-4.0 licence.

give rise to two non-duplicating cancer cells with probability $1 - p$. For simplicity, the rates of duplication and switching are taken to be equal to R_d for all the cells, yielding:

$$\frac{dS}{dt} = R_d(\epsilon S + p(\mu)(C + D)), \tag{5.7}$$

$$\frac{dC}{dt} = R_d(1 - \epsilon)S - C(1 + p(\mu)), \tag{5.8}$$

$$\frac{dD}{dt} = R_d(2C - p(\mu)D). \tag{5.9}$$

When $p = 0$ the model is equivalent to the CSC model introduced in Section 5.2 (La Porta et al., 2012) (in the limit $M = 1$) and discussed above, while the case of constant p has been studied in Zapperi and La Porta (2012). Sellerio et al. (2015) study the case in which the switching probability is not constant but is controlled by the level of relevant miRNAs, which responds to the depletion of the CSC population, as observed experimentally. Assuming that the miRNAs regulating phenotypic switching are produced with a rate that rapidly vanishes when the fraction of CSCs $f_S = S/(S + C + D)$ is sufficiently large, and are cleared at constant rate γ, Sellerio et al. (2015) write

$$\frac{d\mu}{dt} = \beta \exp(-f_s/s_0) - \gamma\mu. \tag{5.10}$$

The switching probability is activated when the level of miRNA is above a threshold μ_0 according to the equation

$$p(\mu) = (1 + \tanh((\mu - \mu_0)/\sigma)). \tag{5.11}$$

Simulations of the model show the presence of an overshoot in the CSC population after sorting (Sellerio et al., 2015). The model illustrates in simple terms how cancer cells react to the depletion of CSCs, showing that phenotypic switching is not just a random event, but an active response to a depletion of the CSC population.

A similar mechanism might be at play in stem cell niches: without switching, the CSC branching process is very similar to stochastic models for stem cell evolution that have been found to describe accurately the distribution of clone size in vivo (Clayton et al., 2007). In the case of stem cells, however, tissue homeostasis requires that, on average, each stem cell gives rise to one stem cell (i.e., the parameter ϵ is equal to zero). In these conditions, the model predicts that the probability of extinction for a clone originating from a single stem-cell tends to one in the long-time limit (Harris, 1989) but when the stem cell pool is large enough, fluctuation-induced extinction is very unlikely (Clayton et al., 2007). If an external perturbation suddenly reduces the stem cell pool, however, the model predicts that extinction becomes likely. Phenotypic switching could thus represent an effective response to this threat.

A direct consequence of these findings is that reducing the number of CSCs below a threshold induces the cells to switch. This implies that a therapeutic strategy based on CSC eradication is unlikely to succeed. Rather than targeting and decreasing the CSCs as the 'root' of the cancer, a balanced reduction in both populations is needed to keep the tumor in a state of asymmetric cell division as it is reduced.

5.4 Cell Sorting and Imperfect Markers

The operative identification of CSCs relies on stem cell markers, but their absolute accuracy is far from being guaranteed. It is indeed reasonable to assume that putative CSC markers are imperfect. An imperfect marker yields a marker-positive subpopulation that is CSC-rich, but does not allow all CSCs to be eliminated from the marker-negative subpopulation. The few CSCs present in the marker-negative subpopulation could drive tumor growth and re-establish a marker-positive subpopulation in the tumor. Furthermore, sorting by FACS typically involves errors: some cells could be assigned to the wrong category. Mathematical models can be used to simulate the effect of imperfect markers and quantify possible sorting errors (Zapperi and La Porta, 2012).

We can define the efficiency of the sorting by η, the probability that a cell is sorted incorrectly by FACS. An imperfect sorting on a cell population characterized by a number $[S^0, C_k^0, D^0]$ of CSCs, CCs and senescent cells yields a positive subpopulation composed by $[(1 - \eta)S^0, \eta C_k^0, \eta D^0)]$ cells and a negative subpopulation composed by $[\eta S^0, (1 - \eta)C_k^0, (1 - \eta)D^0)]$ cells. Consequently the fraction of positive cells in the original population is not equal to the fraction of CSCs but is given by

$$f^{(+)} = (1 - \eta)f_{CSC} + \eta(1 - f_{CSC}), \tag{5.12}$$

and only for $\eta = 0$ do we have $f^{(+)} = f_{CSC}$. Using this model we can study the evolution of positive cells in sorted subpopulations.

To quantify the effect of an imperfect sorting, we can consider the evolution of the concentration of positive cells as a function of the sorting efficiency η. Using the CSC model discussed in Section 5.2, we start from steady-state concentrations of CSCs and CCs and sort them into two subpopulations according to Equation 5.12. Next, we can integrate Equations 5.2 and at each generation compute the fraction of positive cells. The sorting efficiency, while not directly accessible from experiments, can be estimated by a simple calculation. When discussing FACS experiments it is customary to report the purity of the process, obtained by sorting the subpopulation immediately after the first sorting. Here we define the purity κ of the sorting as the real concentration of positive cells present in the nominally positive subpopulation and express it in terms of η

$$\kappa = \frac{(1 - \eta)f_{CSC}}{(1 - \eta)f_{CSC} + \eta f_{CSC}}. \tag{5.13}$$

Combining Equation 5.12 and Equation 5.13, we can estimate the sorting efficiency in terms of the measured values of $f^{(+)}$ and κ. In the limit $f^{(+)} \ll 1$, typical of CSC markers, and high purity $(1-\kappa \ll 1)$, we obtain a simple expression

$$\eta \simeq f^{(+)}(1 - \kappa). \tag{5.14}$$

To make a concrete example, Gupta et al. (2011) reports a purity of 96 percent and 2 percent stem-like cells obtained from SU159 breast cancer cells. Inserting $\kappa = 0.96$ and $f^{(+)} = 0.02$ in Equation 5.14, we estimate $\eta = 8 \times 10^{-4}$.

6

Biomechanics of Cancer

Like any other material, cells respond to mechanical perturbations by deforming, but unlike passive elastic objects, however, they can apply active forces to the environment. Active matter is attracting a large amount of attention in the field of soft matter, with many results having direct relevance to cell mechanics. Here we first introduce a basic notion from the theory of linear elasticity (Section 6.1) and the concept of viscoelasticity. In Section 6.2, we apply these ideas to the mechanics of cancer cells, discussing the relevance of mechanical properties in the malignancy of a cancer cell. In order to grow, a tumor should overcome the mechanical resistance posed by the tissue, as we discuss in Section 6.3. A simple experimental method to study in vitro the effect of stresses on tumor growth relies on osmotic pressure, leading to many quantitative studies that we discuss in Section 6.4. All these different effects are then summarized in Section 6.5, where we review the role of mechanical stresses for cancer progression.

6.1 Elasticity

All materials deform in response to stress, and cells are no exception (Kollmannsberger and Fabry, 2011; Lange and Fabry, 2013). In general, the stiffness of materials is quantified by measuring the Young modulus E, defined as the ratio between the applied stress, the force per unit area, $\sigma = F/A$, and the relative elongation or strain $\epsilon = dL/L$ in the direction of the force

$$\sigma = E\epsilon, \tag{6.1}$$

an equation valid for Hookean solids in the limit of small deformations. In fact, stress and strain are tensorial quantities since one can be interested in forces applied along any directions. Hence, for completeness, we recall here the basic elements of the theory of linear elasticity (Landau and Lifishitz, 1970).

Given the local displacement field, $\mathbf{u}(\mathbf{r})$, it is useful to define the symmetric *strain tensor* as

$$\epsilon_{ik} \equiv \frac{1}{2}\left(\frac{\partial u_i}{\partial x_k} + \frac{\partial u_k}{\partial x_i}\right). \tag{6.2}$$

When a solid deforms, interactions at the molecular scale provide restoring forces that tend to bring the solid back to static equilibrium. Due to the action–reaction principle, the resulting force acting on the volume element V, $\int_V dV \mathbf{F}$, is only given by surface terms. Thus each component of $\mathbf{F}$ is the divergence of a vector field (i.e. $F_i = \sum_k \frac{\partial \sigma_{ik}}{\partial x_k}$) so that

$$\int_V dV\, F_i = \int_S dS \sum_k n_k \sigma_{ik}, \tag{6.3}$$

where S is the surface of the volume V, n_k is a unit vector normal to S and σ_{ik} is the *stress tensor*. The equilibrium condition for the body implies $F_i = 0$, which, in terms of the stress tensor, can be written as

$$\sum_j \frac{\partial \sigma_{ij}}{\partial x_j} = 0. \tag{6.4}$$

In the limit of small deformations, there is a linear relation between stress and strain

$$\sigma_{ij} = \sum_{i,j,k,l} C_{ijkl}\epsilon_{kl}, \tag{6.5}$$

where C_{ijkl} is the elastic moduli tensor. Hooke's law, reported in its most general form in Equation 6.5, usually takes a simpler form since symmetry considerations can be used to reduce the number of independent components of C_{ijkl}. In particular, for isotropic solids, we have only two independent components and the Hooke law can be written as

$$\sigma_{ij} = \frac{E}{(1+v)}\left(\epsilon_{ij} + \frac{v}{1-2v}\delta_{ij}\sum_k \epsilon_{kk}\right), \tag{6.6}$$

where E is the Young modulus and v is the Poisson ratio. For a uniaxial deformation along direction z, Equation 6.6 reduces to $\sigma_{zz} = E\epsilon_{zz}$, which is in the form of Equation 6.1.

In some cases it is convenient to rewrite the strain tensor as the sum of a compressive part, involving a volume change, and a shear part, involving no volume change. Hooke's law then becomes

$$\sigma_{ij} = K\delta_{ij}\sum_k \epsilon_{kk} + 2G\left(\epsilon_{ij} - \frac{1}{3}\delta_{ij}\sum_k \epsilon_{kk}\right), \tag{6.7}$$

where K is the bulk modulus and G is the shear modulus. These coefficients are related to the Young modulus $E \equiv \frac{9KG}{G+3K}$ and the Poisson ratio $\nu \equiv \frac{1}{2}\left(\frac{3K-2G}{3K+G}\right)$. Finally, the equation for stress equilibrium for an isotropic elastic medium can be written in terms of the displacement field **u**, combining Equation 6.6 with Equation 6.4:

$$\frac{E}{2(1+\nu)}\nabla^2\mathbf{u} + \frac{\nu E}{2(1+\nu)(1-2\nu)}\vec{\nabla}(\vec{\nabla}\mathbf{u}) = 0 \qquad (6.8)$$

with appropriate boundary conditions.

6.2 Mechanics of Cancer Cells

Cells are extremely soft materials, with a Young modulus around $E = 1kPA$, which can be measured by different means such as, for example, atomic force microscopy (Moeendarbary et al., 2013). Cells, however, are not simple Hookean solids since their mechanical response is time-dependent and include a viscous component typical of fluids. The simplest description of viscoelasticity is in terms of springs and dashpots (Figure 6.1) that can be combined in series, leading to the Maxwell model, which in its scalar version can be written as

$$\frac{d\epsilon}{dt} = \frac{\sigma}{\eta} + \frac{1}{E}\frac{d\sigma}{dt}, \qquad (6.9)$$

where η is the viscosity, or in parallel, leading to the Kelvin–Voigt model

$$\sigma = E\epsilon + \eta\frac{d\epsilon}{dt}. \qquad (6.10)$$

The Maxwell and Kelvin–Voigt models predict exponential relaxations of strain and stress, respectively, with a characteristic time $\tau - E/\eta$. Relaxation in cells is

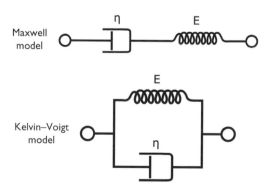

Figure 6.1 Schematic models of viscoelasticity. The Maxwell model considers a spring and a dashpot in series, while in the Kelvin–Voigt model the two elements are in series. Public domain image.

not exponential, as predicted by these models, but decays as a power law (Fabry et al., 2001; Moeendarbary et al., 2013). In particular, the time-dependent response to a constant applied stress σ, or creep, follows

$$\epsilon(t) = \sigma \left(\frac{t}{t_0}\right)^{\beta}, \tag{6.11}$$

with $\beta \simeq 0.1 - 0.5$ (Fabry et al., 2001). Similarly, the stress relaxation in response to a fixed deformation decays as a power law at long timescales

$$\sigma(t) \simeq t^{-\alpha}, \tag{6.12}$$

while at short timescales one observes deviations (Moeendarbary et al., 2013) attributed to the poroelastic behavior of the cell (Mitchison et al., 2008).

Viscoelastic behavior is common to many polymeric materials, but cells are different because their response to stress is not only passive but contains an active component due to acto-myosin-driven contraction of the cytoskeleton. This behavior can be described theoretically by models of active fluids and gels (Liverpool and Marchetti, 2003; Voituriez et al., 2006; Salbreux et al., 2009; Joanny and Prost, 2009; Marchetti et al., 2013; Prost et al., 2015). In the active gel theory, the stress tensor obeys a viscoelastic constitutive equation similar to Equation 6.9 or Equation 6.10 but it is decomposed in the sum of a passive and an active component: $\sigma_{ij} = \sigma_{ij}^{(p)} + \sigma_{ij}^{(a)}$. Here the active stress is written as

$$\sigma_{ij}^{(a)} = \tilde{\zeta}\delta_{ij} + \zeta(\pi_i\pi_j - \delta_{ij}/3) \tag{6.13}$$

where the first term is the "volumetric" stress and the second "deviatoric" term (i.e. not involving volume changes) is coupled to the local orientation π_i of the actin fibers composing the cytoskeleton (Prost et al., 2015). The coefficients $\tilde{\zeta}$ and ζ are proportional to the density of actin fibers and myosin motors.

There is a wide literature on the role of mechanical stresses on tumor growth, as reviewed by Taloni et al. (2015a). The mechanical properties of cancer cells appear to be correlated with their malignancy, in such a way that cells become softer during cancer progression (Fritsch et al., 2010). This has been shown for cell lines (Lekka et al., 1999; Guck et al., 2005) and tumor tissues (Cross et al., 2007; Remmerbach et al., 2009). The distribution of optical deformability of breast tumors shows a distinct shift towards softer cells with respect to normal mammary tissue obtained from surgical breast reductions (Fritsch et al., 2010). The shift is attributed to the change in actin structure from fibrous to diffuse (Sanger, 1975). At large strains, cytoskeletal filaments inherently strain-harden, compensating for the weak linear elastic strength of the actin cortex (Janmey et al., 1991). It has been argued that intermediate filaments such as vimentin could be the right candidate to support the pressure generated by division and dynamics against the surrounding stroma (Chen et al., 2008). It is usually assumed that cell softening is needed by malignant cells to more easily overcome barriers posed by the ECM. The reality is likely more complex since cell migration has been shown to be correlated also with nuclear size, cell adhesion and contractility (Lautscham et al., 2015).

6.3 Tumor Growth Against Tissue-Induced Stresses

Tumors grow at the expense of normal tissues. Hence, space limitations give rise to non-fluid-related pressure, i.e. solid stress. This stress depends on tumor size and local mechanical properties, which are influenced by tumor-associated ECM modifications (degradation, crosslinking, overproduction), altered tissue tensional homeostasis (Bao and Suresh, 2003; Samani et al., 2003; Suresh, 2007), increased compression force due to the solid state pressure exerted by the expanding tumour mass (Paszek and Weaver, 2004), and matrix stiffening due to the desmoplastic response (Paszek et al., 2005). Solid-phase stresses are prominent in elastically stiff regions such as cranium and bones, causing detrimental morbidity. Stresses can be divided into two categories: the externally applied stress, which is developed through mechanical interactions between the solid components of the growing tumor and the surrounding tissue, and the growth-induced stress, which is stored within the tumor as the proliferating cancer, and stromal cells modify the structural components of the tumor microenvironment. The growth-induced stress is a residual stress since it persists even after the tumor is excised and external confining stresses from surrounding tissues have been removed (Skalak et al., 1996; Stylianopoulos et al., 2013, 2012).

At present it is not possible to apply in vitro solid stresses at levels relevant in vivo, and no experimental setup in vitro can faithfully reproduce the heterogeneity that stromal cells display in adult tissues. Nevertheless, in vitro experiments allow the stress exerted at the cellular level and the corresponding long-term dynamics of cell populations to be quantitatively characterized. Furthermore, some in vivo situations may turn out to be very close to in vitro experiments: for instance, growth after surgery. Experiments in vitro have provided the evidence that tumor growth is considerably hindered by the presence of a surrounding constraining medium, such as, for instance, a stiff matrix (Helmlinger et al., 1997; Cheng et al., 2009).

Helmlinger et al. (1997) embedded tumor cells in an inert matrix, where they formed spheroids. The authors found that human colon carcinoma spheroids can grow to a maximum size of 400 mm (diameter) in 0.5% agarose, but only 50 mm in 1.0% agarose (which is less compliant). This was associated with an increase in cell packing and a decrease in cell proliferation. Therefore, a cell confined by a tissue matrix can divide only if its stiffness exceeds the opposing rigidity of its direct environment. Such inhibition of tumor growth can be reversed by releasing the

spheroids from the gel. Moreover, by culturing highly metastatic rat prostate carcinoma (AT3) and the low metastatic variant (AT2) in the same agarose gel matrix
used by Helmlinger et al. (1997), Koike et al. (2002) showed that stress facilitates the formation of spheroids in metastatic lines, while AT2 cells maintained
their spheroidal morphology even after stress removal. These observations have
raised the interesting possibility that a stiffer matrix can be viewed as a host barrier
to tumor invasion, and increasing tissue stiffening could actually impede cancer
progression; an hypothesis corroborated by further recent studies with the help of
different techniques. In 2009, Cheng et al. (2009) showed that the accumulating
solid stress fields in agarose gels around growing tumor spheroids (non-metastatic
murine mammary carcinoma 67NR) dictate the shape of the spheroids and suppress cell proliferations, inducing apoptosis in regions of high solid stress (see
Figure 6.2).

While the work discussed above suggests that stress hinders tumor growth, other
past studies have suggested that changes in ECM structure or mechanics, such

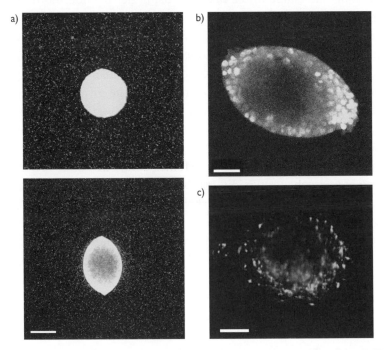

Figure 6.2 Effect of confining stress on the growth of tumor spheroids. a) Tumor
spheroids can fracture the surrounding medium, turning into an oblate shape.
Microbeads are used to quantify the strain field. b) Proliferating cells accumulate in the regions of minimal stress. c) Necrosis accumulates especially in highly
stressed regions. Scale bars are 50μm. Adapted from images by Cheng et al.,
2009, CC-BY-3.0 licence.

as whether the matrix is stiff enough to resist cell traction forces, might actively contribute to tumor formation (Ingber et al., 1981). The reasons for this apparent discrepancy may include the dynamic nature of stress, deficient instrumentation for measuring it in vivo, and challenges in evaluating its individual biological role. Another possible reason is that mechanical stresses displace the homeostasis threshold by inhibiting simultaneously proliferation and apoptosis rates, also inducing phenotypic changes.

6.4 Effect of Osmotic Pressure on Cancer Cells

An alternative method to study the role of mechanics in cancer cell growth is to use osmotic pressure, the pressure exerted on a semipermeable membrane by a solution. While osmotic pressure has a different origin than a mechanical one, the effect is similar: as discussed in 1901 by J. van't Hoff in his Nobel lecture, the osmotic pressure p arises as a result of the collisions of the dissolved molecules with the semi-permeable membrane. In the case of cells, molecules exert a mechanical pressure on the cell membrane leading to a deformation in the cell. Figure 6.3 reports an illustration of this effect for red blood cells placed in a hypertonic (cells shrink), isotonic (cells are in equilibrium with the medium) and hypotonic (cells swell) medium. Hence, there is no fundamental difference with the effect of an equivalent solid stress.

Montel et al. (2011) used a solution of the biocompatible polymer dextran to induce a constant pressure on carcinoma cell (CT26) spheroids. Controlling osmotic pressure allowed the monitoring with unprecedented accuracy of the actual stress exerted at the cell level (Montel et al., 2011, 2012), showing that tumor spheroids reached a steady state corresponding to a typical diameter of 900 μm, corresponding to the establishment of an homeostatic balance between the apoptotic core and proliferating rim (Freyer and Sutherland, 1986; Casciari et al., 1992). Using a similar method, Taloni et al. (2014) studied the proliferation of two melanoma cell lines, primary (IgR39) and metastatic (IgR37), when subjected to dextran-induced osmotic pressure. The results showed that a constant low osmotic pressure affects considerably the proliferation of primary tumors if compared to the corresponding metastatic ones (see Figure 6.4). The same experimental setup allowed the effect of pressure on cell motility and transmigration capability to be studied (La Porta et al., 2015b). The general conclusion is that pressure affects the functional properties of primary melanoma cells, but much less those of metastatic cells. This result suggests the idea that mechanical stress acts as a selective pressure towards more metastatic cancer phenotype.

Hypertonic   Isotonic   Hypotonic

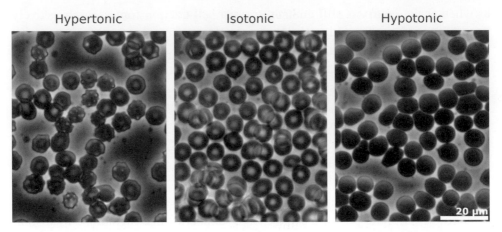

Figure 6.3 Effect of osmotic on red blood cells. Cells swell when placed in an isotonic solution and crumple in a hypertonic solution. Image by Richard Wheeler,
CC BY-SA 3.0 license.

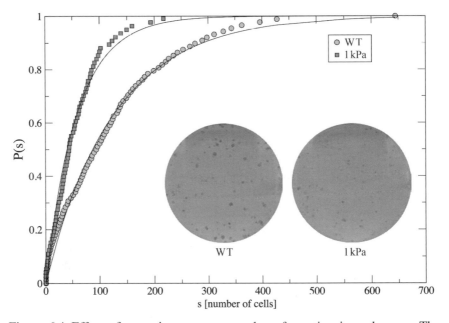

Figure 6.4 Effect of osmotic pressure on colony formation in melanoma. The
cumulative distributions of colony size obtained from IgR39 cells under 1 kPa
osmotic pressure with respect to the control (0 kPa). The curves fit with a continuous time branching process model (see Baraldi et al., 2013), yielding a rate
division of $\gamma = 0.60$ that is reduced to $\gamma = 0.51$ under osmotic pressure. The
images show two representative examples of the colonies for 0 kPa and 1 kPa
conditions. Adapted from an image by Taloni et al., 2014, CC-BY-4.0 licence.

6.5 Mechanical Stresses and Cancer Progression

The tumor-associated stroma is composed of several cell types, including adipocytes, fibroblasts, and immune cells, as well as a variety of ECM proteins, such as collagen and fibronectin. The onset and progression of cancer leads to distinct changes in the stiffness of the ECM surrounding tissue. Indeed, tumors are often detected as a palpable stiffening of the tissue (Huang and Ingber, 2005), and approaches such as magnetic resonance imaging elastography and sono-elastography have been developed to exploit this observation to enhance cancer detection (Glaser et al., 2006; Garra, 2007). Furthermore, altered stromal–epithelial interactions precede and could even contribute to the malignant transformation (Levental et al., 2009). In fact, the desmoplastic stroma that is present in many solid tumors is typically significantly stiffer than normal (Paszek et al., 2005). The cause of the increase in tumor stromal stiffness and its relation with cancer progression are still open questions in a mechanical view of tumorigenesis.

Mechanical stresses are perceived and integrated by the cell at the molecular level through mechanically responsive sensors triggering biochemical signaling cascades which yield a specific cellular response, a process known as *mechanotransduction*. Once mechanical cues have been detected, indeed, cells must propagate, amplify the physical cue created in each of them and translate the signal into either a transient response or a sustained cellular behavior. For instance, force and growth factor receptors can each influence cell growth, survival, differentiation, shape and gene expression by regulating the activity of RhoGTPases that, in turn, modulate actomyosin contractility and actin dynamics (Gieni and Hendzel, 2008; Engler et al., 2004, 2006; Georges et al., 2006; McBeath et al., 2004; Chrzanowska-Wodnicka and Burridge, 1996; Paszek et al., 2005; Whitehead et al., 2008). Another example of mechanotransducer is integrins which interact with the ECM (ECM) (Ginsberg et al., 1992; Hynes, 1992). They regulate integrin-dependent extracellular-signal-regulated kinase (Erk) that is involved in cell motility affecting the capability of the cells to interact with the environment through the assembly of focal adhesion (Lee and Nikodem, 2004; Vial et al., 2003; Levental et al., 2009). All this evidence shows how the cells respond actively to solid stresses altering biochemical pathways and finally remodeling the cytoskeleton. Paszek and Weaver (2004) hypothesized that cellular biochemical cues, such as ERK-MAPK signaling regulating Rho and Rac activity and cytoskeleton contraction observed in colon carcinoma cells (Vial et al., 2003), could be triggered by a compromised tensional-homeostasis, or, in other words, by a stiff environment.

As our understanding of the relation between stromal collagen density, matrix stiffness and tumor progression improves, the underlying molecular mechanisms are slowly unraveling. Wells (2005) showed that the growth factor TGF-β (acting

through the cytoplasmic signaling intermediate Smad3) and the mechanical properties of the underlying matrix play particularly important roles in hepatic stellate cell
transdifferentiation, which is the primary cause of liver fibrosis. Whitehead et al.
(2008) demonstrates that mechanical strain, such as that associated with intestinal
transit or tumor growth, can be interpreted by cells of preneoplastic colon tissue as
a signal to initiate a β-catenin dependent transcriptional program. Levental et al.
(2009) contribute further insights by establishing a correlation between collagen
crosslinking and matrix stiffness in vivo and by implicating the ECM-crosslinking
enzyme lysyl oxidase (LOX) as a culprit driving stiffness-associated tumor progression. Demou (2010) introduced two novel systems, the cell-pressors, to enable
molecular analyses and live imaging of 3D cell cultures under compression. This
study demonstrates for the first time that normal loading could regulate expression
of genes involved in ECM degradation, cell–cell contact, migration and proliferation. The effect of osmotic pressure on the activity of MAPKs and on signal
transduction via the RTK PDGF receptor, which plays an important role in cell proliferation, cell survival, cell migration, and tissue homeostasis, has been reported
in Nielsen et al. (2008).

    High mechanical forces could compromise cell differentiation and alter
mechano-responsiveness, thus promoting the malignant transformation. This is not,
however, a one-way process. Cell traction forces generated in the actin cytoskeleton
are exerted on these same sites so that integrins and focal adhesions are maintained in a state of isometric tension. A rise of cell tension further increases ECM
stiffness by tensing or realigning ECM components, thereby creating a deadly,
self-sustaining positive feedback loop: cells generate more force, disrupt cell–
cell junctions, spread, increase proliferation, and lose acinar organization. Indeed
Gordon et al. (2003) used a three-dimensional, Matrigel-based, in vitro assay, containing $1.0\ \mu$m latex beads, to probe the environment of a dynamically expanding,
multicellular brain tumor spheroid. This study showed that an expanding microscopic tumor system exerts both significant mechanical pressure and significant
traction on its microenvironment. Paszek et al. (2005) explored the role of bidirectional force transfer across integrins in the context of differentiation and tumor
formation. Using an electromechanical indentor they showed that explanted mouse
mammary tumors are stiffer than healthy mammary glands. Moreover, culturing
normal mammary epithelial cells on ECM gels with varying mechanical compliance, they observed that stiff (force-resisting) ECM gels promote expression of the
undifferentiated malignant phenotype, and Rho activity was also higher in these
cells. Using both laser-scanning multiphoton and second harmonic generation
microscopy, Provenzano et al. (2006) detected local alterations in collagen density around tumors. Moreover, local cell invasion was found predominantly to be
oriented along certain aligned collagen fibers, suggesting that radial alignment of

collagen fibers relative to tumors facilitates invasion. Importantly, regions of high breast density were associated with increased stromal collagen (Provenzano et al., 2008). Indeed, an increase in ECM protein concentration, i.e. an increased matrix crosslinking or parallel reorientation of matrix fibrils within a stromal matrix, can stiffen a tissue locally to alter cell growth or direct cell migration, albeit to differing degrees.

Other in vitro experiments validate those findings: by exerting compressive stress by means of a weighted piston in 2D assays containing mammary epithelial cell lines, Janet et al. (2012) showed that the compressive stress accumulated during tumor growth can enable coordinated migration of cancer cells, by stimulating formation of leader cells and enhancing cell-substrate adhesion. More recently, Alessandri et al. (2013) developed a new method based on the encapsulation and growth of cells inside permeable, elastic, hollow microspheres. These confining conditions are observed to increase the cellular density and affect the cellular organization of the spheroid, while, performing invasion assays in a collagen matrix, the peripheral cells readily escape preconfined spheroids. These results suggest that mechanical cues from the surrounding microenvironment may trigger cell invasion from a growing tumor.

7

Cancer Cell Migration

Cell migration plays a fundamental role in cancer, being at the basis of the metastatic process, where cancer cells detach from the primary tumor, invade neighboring tissues and finally spread to distant organs. In this chapter we discuss the main biological and physical aspects of migration, starting from the motion of individual cells (Section 7.1) and the statistical characterization of their trajectories in terms of stochastic processes. Cells respond to chemical signals, in a process known as *chemotaxis* (Section 7.2), and then polarize and move in the direction of the chemical gradient. In order to move, cells make use of a vast class of cell adhesion molecules, which we review in Section 7.3. Motion results from the application of traction forces to the ECM, which can be quantified experimentally, as we discuss in Section 7.4. Cell adhesion molecules also play an important role in the interaction between cells, leading to collective effects. Experiments and models of collective cell migration are discussed in Section 7.5. Finally, in Section 7.6, all the fundamental steps giving rise to cell migration are summarized and discussed in the framework of the metastatic process.

7.1 Individual Cell Motion

Cell migration plays a key role in many physiological processes such as embryogenesis and morphogenesis, immune response, wound healing and tissue repair, but it is also a crucial determinant for cancer invasion and metastasis (Rørth, 2009; Friedl and Gilmour, 2009; Ilina and Friedl, 2009). Cellular migration is regulated at the biochemical level by the coordinated action of a multitude of factors involved in the response to external chemical stimuli, in the remodeling of the actin cytoskeleton and in the regulation of adhesion molecules needed to apply traction forces to the surrounding ECM and to neighboring cells. While discussing these processes in more detail in the coming sections, we focus here on the movement resulting from their integration.

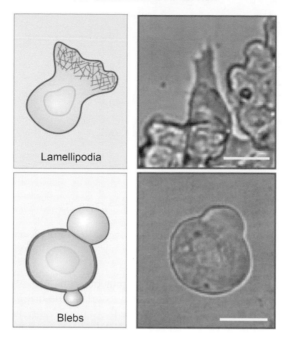

Figure 7.1 The two main modes of cell migration are based on the formation of lamellipodia or blebs. The images refer to zebrafish primordial germ cells. Adapted from an image by Ruprecht et al., 2015, CC-BY-4.0 licence.

Broadly speaking, we can identify two main migration modes for individual cells (Ruprecht et al., 2015): i) Lamellipodia-driven mobility, which relies on the formation of a single lamellipodium at the cell leading edge, and ii) amoeboid cell migration, which is based on the formation of blebs in the plasma membrane (Figure 7.1) (Fackler and Grosse, 2008; Paluch and Raz, 2013).

Lamellipodia are large, thin sheets formed at the leading edge of the cell by an actin mesh. Motion is accomplished by a combination of actin turnover and adhesion to the substrate. Actin filaments typically polymerize at the front and depolymerize at the back, resulting in a treadmilling movement. Amoeboid cell migration is due instead to the formation of blebs, round protrusions in the plasma membrane due to its detachment from a contracting acto-myosin cortex (Charras et al., 2005, 2006). A crucial role in bleb expansion is played by water transfer through the membrane, thanks to the the aquaporin water channel (Taloni et al., 2015b). This is also reflected by the fundamental importance of aquaporin in affecting the migration properties of cancer cells (Monzani et al., 2009). In confined environments, cells are able to migrate relying exclusively on water transport even when actin polymerization and myosin driven contraction are inhibited (Stroka et al., 2014).

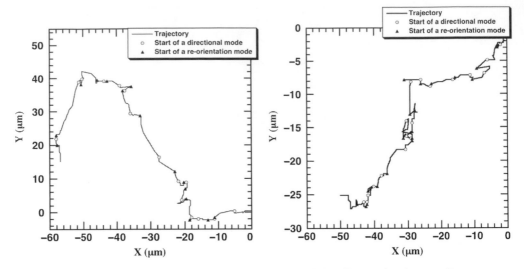

Figure 7.2 Two trajectories of mammary epithelial cells moving in two dimensions for two hours. The cell trajectory always starts at the origin. An open circle indicates the start of directional motion while the start of re-orientation is denoted by a filled triangle. Image by Potdar et al., 2010, CC-BY-4.0 licence.

The simplest method to investigate individual cell migration is to observe cell trajectories *in vitro*, as those reported in Figure 7.2 (Potdar et al., 2010). The trajectories display an erratic behavior typical of Brownian motion (Codling et al., 2008), but a closer quantitative analysis shows that cells do not perform a simple random walk (Stokes et al., 1991; Dieterich et al., 2008; Potdar et al., 2010; Wu et al., 2014). In particular, Figure 7.2 shows periods of persistent directional motion in one direction followed by reorientation events when the direction changes completely. The simplest stochastic process that reproduces this behavior is the persistent random walk (PRW) described by the Langevin equation (Stokes et al., 1991; Wu et al., 2014):

$$\frac{d\mathbf{v}}{dt} = -\frac{\mathbf{v}}{\tau} + \sqrt{\frac{D}{\tau}}\eta(t), \tag{7.1}$$

where $\mathbf{v}$ is the cell velocity, τ is the persistence time, $\eta(t)$ is an uncorrelated Gaussian noise with zero mean and unit variance, and the noise strength is tuned by D. This linear stochastic model can be easily solved and yields a mean-square displacement

$$\langle (\mathbf{r}(t + t_0) - \mathbf{r}(t_0))^2 \rangle = 2D\tau \left(\exp(-t/\tau) - \frac{t}{\tau} - 1 \right). \tag{7.2}$$

Equation 7.2 interpolates from an exponential increase at short times to a linear diffusive behavior at large times, which agrees with experimental data for two

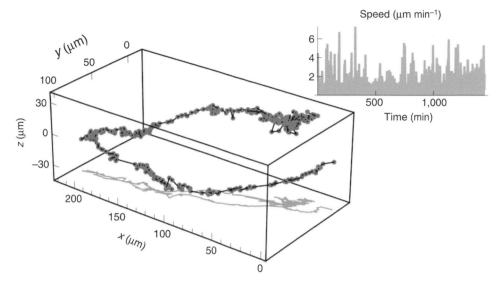

Figure 7.3 A trajectory of a breast carcinoma cell migrating in a three-dimensional collagen matrix. The inset shows the intermittent fluctuations of the cell velocity. Image by Metzner et al., 2015, CC-BY-4.0 licence.

dimensional motion (Wu et al., 2014). An alternative model to explain the deviation from pure Brownian motion involves invoking anomalous diffusion (e.g., a mean square displacement scaling as $t^{2\alpha}$, with $\alpha > 1/2$) instead of persistence (Dieterich et al., 2008).

Recent experiments tracked cell motion in three-dimensional collagen matrices (see Figure 7.3), revealing clear deviation from the simple PRW model (Wu et al., 2014; Metzner et al., 2015). In particular, experiments show that the distribution of velocities is not Gaussian, as predicted by the PRW (Wu et al., 2014). The correct distribution can be obtained by modeling cell heterogeneity and substrate anisotropies (Wu et al., 2014) and, more recently, by introducing a superstatistical framework (Metzner et al., 2015) where the motion is modeled by a persistent random walk with parameters (e.g., τ and D) that are themselves evolving stochastically. The superstatistical framework allows excellent results to be obtained for cell motion in two and three dimensions (Metzner et al., 2015).

7.2 Chemotaxis

Cells move in response to external stimuli and, in particular, to changes in the chemical environment, a process known as *chemotaxis* (see Figure 7.4). Contrary to bacteria, eukaryotic cells are large enough to be able to sense chemical gradients and respond to them by polarizing their cytoskeleton, moving towards the

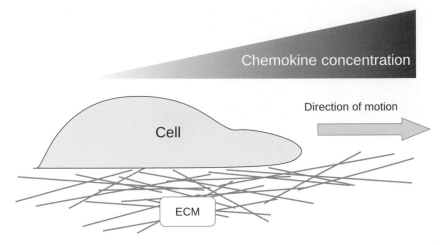

Figure 7.4 Schematic of chemotaxis. Cells sense concentration gradients and move towards the larger concentrations.

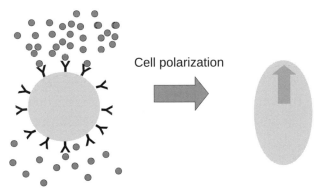

Figure 7.5 Surface receptor molecules are spread uniformly on the surface of the cell. In the presence of a chemical gradient an imbalance in the number of bound receptors from top to bottom leads to the polarization of the cell towards the top direction.

higher concentration (Weiner, 2002; Van Haastert and Veltman, 2007; Iglesias and Devreotes, 2008). This is accomplished through a uniform array of receptor molecules sitting on the cell surface that bind specific chemoattractant molecules (see Figure 7.5), such as chemokines, which will be discussed in more detail in Section 9.3. Receptor binding leads to a signal transduction inside the cell through the activation of complex pathways. The cell must then be able to detect differences in signaling resulting from differences in occupied receptors on the surface, a process referred to as "gradient sensing" (Parent and Devreotes, 1999). As a result of this, cells polarize, changing their cytoskeleton by actin polymerization, forming pseudopodia in the front of the cell and finally moving in the direction of the higher

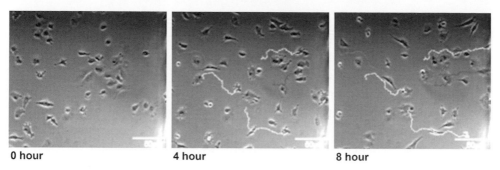

0 hour   4 hour   8 hour

Figure 7.6 An example of chemotaxis of melanoma cells in a gradient of concentration of Fetal Bovine Serum. Images are taken after 4 or 8 hours. Higher concentrations are in the right part of the image. Cell tracking lines show the trajectories of the cells in the direction of the higher concentration. Image by Muinonen-Martin et al., 2010, CC-BY-4.0 licence.

gradient. An example of melanoma cells moving in response to a chemical gradient is shown in Figure 7.6.

It is important to remark that gradient sensing does not necessarily imply movement. Evidence shows that cells are able to sense gradients even when they do not move (Parent and Devreotes, 1999). The gradient sensitivity in some eukaryotic cells can be extremely precise: experiments show that cells can detect concentration differences that are below 1 percent (Rosoff et al., 2004). How such a sensitive detection can be achieved in the presence of noise is a complex and intriguing problem (Herzmark et al., 2007; Rappel and Levine, 2008b,a).

A mean-field kinetic description of gradient sensing can be based on a simple ligand-receptor binding reactions $L + R \rightleftharpoons LR$ for the front and back of a cell placed in a concentration gradient p such that $L_f = c(1 + p)$ and $L_b = c(1 - p)$ (Herzmark et al., 2007). The kinetics reaction can be written as (see Section 3.3.1)

$$\frac{d[L_f]}{dt} = k_+[L_f][R_f] - k_-[LR_f] \tag{7.3}$$

$$\frac{d[L_b]}{dt} = k_+[L_b][R_b] - k_-[LR_b], \tag{7.4}$$

which in the steady state ($t \to \infty$) yields signals S as the fraction of occupied receptors (Rappel and Levine, 2008a)

$$S_f = \frac{[LR_f]}{[LR_f] + [R_f]} = \frac{c(1 + p)}{c(1 + p) + K_d} \tag{7.5}$$

$$S_b = \frac{[LR_b]}{[LR_b] + [R_b]} = \frac{c(1 - p)}{c(1 - p) + K_d}, \tag{7.6}$$

where $K_d = k_+/k_-$ is the dissociation constant. These equations are fully deterministic and predict a signal difference between front and back $S_f - S_b$ that is proportional to p at small gradients. In general, however, one should consider an additional noise term to Equations 7.5 and 7.6. When the signal to noise ratio is too small it would be impossible to detect the gradient. Hence the cell must possess an internal mechanism to reduce the noise.

Levine et al. (2006) propose a model based on balanced inactivation following the general concept of excitation-inhibition in which an occupied receptor triggers two contrary responses: a rapidly increasing excitation and a slowly increasing inhibition. The combination of these two processes can lead to complex responses like a time overshoot followed by a plateau (Parent and Devreotes, 1999). Levine et al. (2006) postulate that the inhibitor concentration is locally varying and follows the diffusion equation:

$$\frac{\partial[B]}{\partial t} = D\nabla^2[B]. \tag{7.7}$$

The inhibitor field B is produced with a rate-proportional S, binds with the activator field A, also produced with a rate proportional to the signal S, and can attach to the membrane. The concentration of the inhibitor field at the membrane B_m follows

$$\frac{\partial[B_m]}{\partial t} = k_b[B] - k_{-b}[B_m] - k_i[A][B_m] \tag{7.8}$$

and the activator field evolves according to

$$\frac{\partial[A]}{\partial t} = k_a S - k_{-a}[A] - k_i[A][B_m], \tag{7.9}$$

where k_a, k_b, k_{-a}, k_{-b} and k_i are reaction fields. Finally the evolution equations have to be complemented with a boundary condition at the membrane

$$\frac{\partial[B]}{\partial n} = k_b S - k_b[B], \tag{7.10}$$

where n is the normal to the membrane. The analysis of the model shows that the noise in external signal is effectively integrated, leading to a switch-like response (Levine et al., 2006; Rappel and Levine, 2008a).

7.3 Cell Adhesion Molecules

Migration requires that individual cells attach to the surrounding ECM, while collective cell migration also involves attachments to neighboring cells. At the microscopic level, cell–cell and cell–ECM attachments are possible thanks to cell adhesion molecules (CAM). Through their fundamental role in cell–cell adhesion and in anchoring cells to the ECM, CAM provide structure to cells and tissues, allow cell signalling, and promote tissue repair and wound healing (Makrilia et al., 2009). As illustrated in Figure 7.7, the most important classes of CAM are

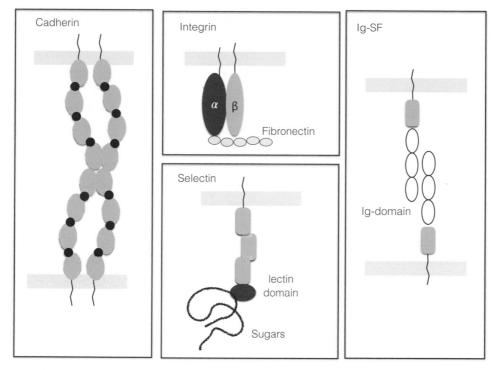

Figure 7.7 Schematic of the most important classes of cell adhesion molecules. The diagram illustrates the type of adhesion mechanisms for these proteins. Cadherins form bonds with similar molecules of other cells. Integrins attach to ECM fibers such as fibronectin. Selectin binds carbohydrates and Ig-SF molecules bind through their Ig-domain.

cadherins, integrins, selectins and members of the immunoglobulin superfamily (IgSF). More specifically, the main role of cadherins is to insure cell–cell adhesion, playing a critical role in the epithelial-mesenchymal transition, in cell migration and in gene regulation through catenins, namely by the β-catenin/Wnt signaling pathway (Makrilia et al., 2009). Integrins, on the other hand, are involved in both cell–cell and cell–ECM interactions and they play a crucial role in cell proliferation, differentiation and migration due to their ability to transfer signals from the ECM into the cell (Giancotti and Ruoslahti, 1999). Selectins and IgSF family members are mainly involved in wound healing and in the immune response. In particular, these molecules act as "traffic regulators" inside the lymphatic system by attracting immuno-competent cells to the inflammation site (Borsig et al., 2002).

Integrins are α/β heterodimeric transmembrane receptors which in addition to cell adhesion regulate various cellular responses promoting proliferation, migration and survival. Integrins can directly and indirectly recruit a number of signaling

components and activate intracellular signaling cascades, especially pathways leading to the activation of MAPK pathway. Cells typically express a variety of
different integrins which can bind to many different ligands and in particular to
RGD (arginine-glycine-aspartate) amino acid sequences that are found in ECM
components such as collagen, fibronectin or laminin. The cytoplasmic domain
of the integrin molecule connects to the cytoskeleton by linking actin filaments
through other proteins such as talin, vinculin, or α-actinin. In this way, integrins
represent a very effective tool for anchoring the cell to the ECM.

Selectins are single-chain transmembrane glycoproteins whose transmembrane
lectin domain can bind carbohydrate groups (Figure 7.8). Cell–cell adhesion is
possible by binding selectin molecules from one cell to carbohydrates on the surface of neighboring cells. We can distinguish three types of selectins: E-selectins,
expressed by epithelial cells in response to cytokine stimulation; P-selectins, found
in blood platelets; and L-selectins, expressed by leukocytes. Selectin adhesion is
enhanced by shear, as observed in leukocytes that in the blood stream adhere to
endotelial cells only when the shear stress is larger than a threshold. Furthermore,
the number of adherent leukocytes increases with shear (Ramachandran et al.,
2004). This counter-intuitive behavior can be explained, assuming that selectin
binding occcurs in the form of a *catch-bond* (Thomas, 2008). While a normal
slip-bond weakens under tension and therefore its lifetime decreases, a catch-bond
becomes stronger and its lifetime increases. Catch-bonds are often encountered
in biology and have the function to stabilize bonds under tension (Thomas et al.,
2008).

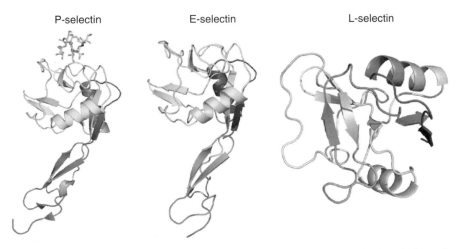

Figure 7.8 The structure of the three selectin proteins. P-selectin (rendering of
PDB 1G1R, by C. Neveu), E-selectin (rendering of PDB 1esel, by Emw), and
L-selectin (rendering of PDB 1KJB). Images CC-BY-SA-3.0 licence.

Cadherins are calcium-dependent, type 1 transmembrane proteins and can be classified into four groups: classical, desmosomal, protocadherins, and unconventional. For instance, E-cadherin, expressed in most epithelial cells, and N-cadherin, expressed in neural tissues, both belong to the classical group. Cadherin extracellular domains display repetitive subdomains that can bind calcium, leading to conformational changes that allow the formation of bonds with cadherin molecules of neighboring cells. Single-molecule force measurements have shown that these bonds become stronger under mechanical forces (i.e., they behave as catch bonds) (Rakshit et al., 2012). Further experiments and simulations indicate that the catch bond is due to the formation of hydrogen bonds that lock cadherin extracellular domains and that the process is regulated by Ca2+ ions (Manibog et al., 2014). The cytoplasmic domain of cadherin instead binds to the cytoskeleton through intermediate proteins (Figure 7.9). In addition to physically connecting neighboring cells, cadherins are also involved in signal transduction through β-catenin, a multifunctional protein that plays a major role in embryonic development, organogenesis and cellular homeostasis. In the absence of Wnt signaling, β-catenin is bound in a complex with axin, APC and GSK3-β and is constantly degraded through phosphorylated and ubiquitin-mediated proteasomal degradation. In the presence of Wnt signaling, the binding complex dissociates

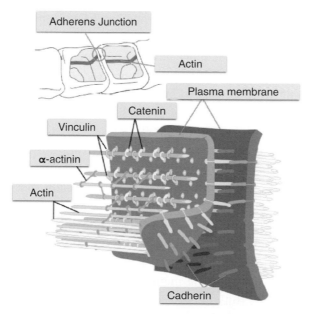

Figure 7.9 Schematic of cadherin-based adherens junction. Actin filaments are linked to α-actinin and to membrane through vinculin. The head domain of vinculin associates to E-cadherin via catenins while its tail domain binds to membrane lipids and to actin filaments. Image by M. Ruiz, public domain.

and β-catenin translocates to the nucleus where it binds T-cell factor/lymphoid enhancer factor (LEF) transcription factors, activating target genes involved in cell proliferation and cell adhesion, i.e., matrix metalloproteases (Wodarz and Nusse, 1998; Huelsken and Birchmeier, 2001). Furthermore, Wnt-signalling inhibits the degradation of β-catenin by the transmission of signal from the disheveled protein (DSH) GSK-3β. The main role of GSK-3β is to phosphorylate β-catenin within a complex with APC/axin and trigger its ubiquitination and degradation. β-catenin is also found in a cadherin-bound form at the plasma membrane where it is required for the formation and stabilization of the adherens junction in order to support proper tissue architecture and morphogenesis (Gumbiner, 2000; Kam and Quaranta, 2009).

The immunoglobulin superfamily (IgSF) is a large class of proteins that share key structural features with immunoglobulins. In particular, they all display immunoglobulin domains (Ig-domain) consisting of two layers of antiparallel β-strands arranged in two β-sheets. IgSF molecules are mostly involved in the immune system, acting as antigen receptors, but they also work as adhesion molecules by binding integrins or other IgSF molecules from other cells.

7.4 Traction Forces During Cancer Cell Migration

As discussed above, in order to migrate and invade neighboring tissues, cancer cells have to exert mechanical forces on the surrounding ECM (Lange and Fabry, 2013). Understanding this point is important since experiments revealed significant differences in traction forces between metastatic and non-metastatic cells (Kraning-Rush et al., 2012). Cellular traction forces are, however, not only involved in cancer metastasis (Kraning-Rush et al., 2012) and angiogenesis (Califano and Reinhart-King, 2010), but also in physiological processes such as contraction (Luna et al., 2012), cell migration (Oakes et al., 2009); Théry (2010); Ricart et al. (2011), and wound healing (Trepat et al., 2009; Brugues et al., 2014). Hence, a precise quantification of traction forces is becoming an extremely important challenge.

In two dimensions, cellular traction forces are typically measured by placing cells on a planar flexible gel substrate on which beads are embedded. The strain field on the substrate is reconstructed by observing the displacements of the beads after the cells are removed (see Figure 7.10). Traction forces are then obtained by inverting the relationship between displacements and tractions on an elastic substrate of known elasticity using mathematical algorithms (Harris et al., 1980; Dembo and Wang, 1999; Butler et al., 2002; Merkel et al., 2007; Liu et al., 2013).

To be more specific, if an elastic substrate is subjected to a surface force field with components $f_i(\mathbf{r})$, its displacement field is given by

$$u_j(\mathbf{r}) = \sum_{i=1}^{3} \int d^3r' G_{ij}(\mathbf{r}' - \mathbf{r}) f_i(\mathbf{r}'), \qquad (7.11)$$

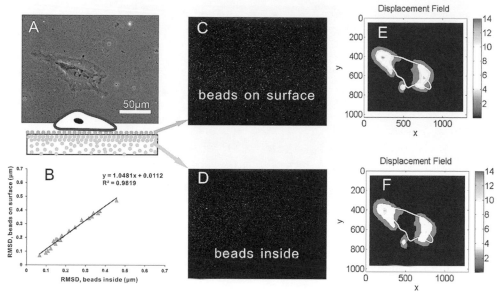

Figure 7.10 Two-dimensional traction-force microscopy. A) Phase-contrast image of a HeLa cancer cell on a gel substrate. Beads are mixed inside and linked on surface of the gels, as illustrated in the inset. B) Scatter plot of bead root mean-square displacement (RMSD) computed for 22 cells utilizing fluorescence images of the beads on surface (vertical axis) and beads inside (horizontal axis). C) Fluorescence image of beads on surface. D) Fluorescence image of beads inside gels. E) Displacement field was calculated using the fluorescence image of beads on surface before and after cell removal. F) Displacement field calculated using the fluorescence image of beads inside before and after cell removal. The solid white line stands for the cell outline in both E) and F). Image by Liu et al., 2013, CC-BY-4.0 licence.

where G_{ij} is the elastic Green tensor (Landau and Lifishitz, 1970). The displacement field $u_j(\mathbf{r})$ in the substrate can be estimated by digital image correlation processing from the position of the embedded beads (see Figure 7.11) and the traction force field $f_i(\mathbf{r})$ is obtained by inverting Equation 7.11. Inversion can easily be done using Fourier transforms (Butler et al., 2002) or by other conventional matrix inversion methods (Dembo and Wang, 1999; Merkel et al., 2007).

While two-dimensional traction force measurements are widely used as an in vitro assay for cancer cell mechanics, tumor invasion occurs in three-dimensional tissue. It is, however, possible to extend the method by dispersing beads in three dimensional cell cultures. Using this approach, it is possible to measure the distribution of cell tractions using a synthetic polymer gel with linear elastic properties as a matrix (Legant et al., 2010). More recently the methodology was improved by computing the elastic strain energy stored in the matrix due to traction-induced deformations (Koch et al., 2012). An example of the strain energy map in three dimensions is shown in Figure 7.12. Using this approach, it was shown that

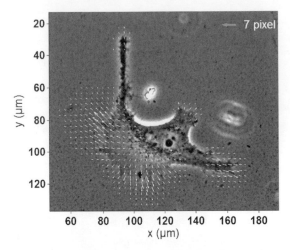

Figure 7.11 Actual displacement fields from traction-force microscopy after corrections by an image processing algorithm. Image by Liu et al., 2013, CC-BY-4.0 licence.

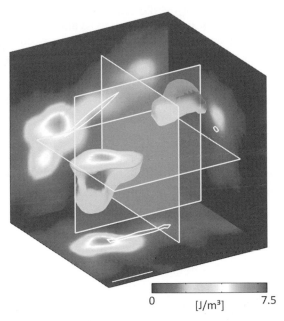

Figure 7.12 Strain energy density around an elongated MDA-MB-231 breast carcinoma cell embedded in a collagen gel. The strain energy density is highest in close proximity to the cell poles and decays rapidly in all directions away from the cell. Image by Koch et al., 2012, CC-BY-4.0 licence.

invasive breast and lung carcinoma cells displayed a higher contractility than non-invasive cells, but this signature was not universal. The cell morphology of invasive cells during migration appeared to be universally more elongated than that of non-invasive cells (Koch et al., 2012).

7.5 Collective Cell Migration

Cancer cells often do not invade neighboring tissues alone but move collectively (Rørth, 2009; Khalil and Friedl, 2010; Gov, 2014). Collective cell migration is a key feature of epithelial tissue but it is often encountered in cancer metastasis. This observation is at odds with the conventional idea according to which cancer cells first undergo an epithelial-to-mesenchimal transition and then migrate individually. While in individual cell migration chemotaxis, mechanotransductions and the resulting signaling cascades all occur within a single cell, in collective cell migration all these functions are coherently shared by a group of cells that can behave as a multicellular unit in which internal communication takes place via cell–cell junctions.

A widely used experimental technique to quantify collective cell migration is the wound-healing assay in which a confluent cell monolayer is scratched and the motion of the ensuing cell front is observed by time-lapse microscopy as it invades the empty space. Traction-force microscopy during wound healing reveals that collective motion arises from mechanical stresses transmitted between neighboring cells (Tambe et al., 2011), giving rise to long-ranged stress waves in the monolayer (Serra-Picamal et al., 2012; Banerjee et al., 2015). Experiments performed on plates with different coating such as gels (Haga et al., 2005; Ng et al., 2012), micro-patterned (Röttgermann et al., 2014) or deformable substrates (Tambe et al., 2011) show that cell migration is guided by the substrate structure and stiffness (Oakes et al., 2009; Angelini et al., 2010; Koch et al., 2012; Metzner et al., 2015). Bazellières et al. (2015) have investigated the precise role of cell–cell adhesion for collective migration by knocking down more than twenty individual CAMs. The results show that P-cadherin is related to the strength of the adhesion force, while E-cadherin controls the rate of its buildup. Even when a wound is closed, cells do not stop moving but slowly rearrange in a way that resembles atomic relaxation in glasses (Angelini et al., 2011). Depending on the cell shapes and adhesion strength, cell motion can be either jammed or flowing in close analogy with what is observed in conventional glassy systems (Park et al., 2015).

The vast experimental literature on collective cell migration is complemented by several computational and theoretical models in the broader field of active matter (Vedula et al., 2013; Szabó et al., 2006; Poujade et al., 2007; Sepúlveda et al., 2013). In these models, cells are a set of self-propelled interacting particles

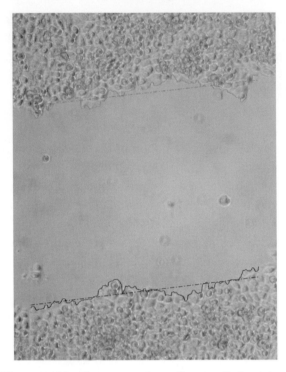

Figure 7.13 The wound healing assay. A confluent cell sheet, in this case composed of HeLa cells, is scratched and the closing motion of the front is observed in time-lapse microscopy. Here, the front profile and its average position are displayed by solid and dashed lines, respectively.

and analyzes their collective motion. Interactions typically include attraction, due to adhesion, between neighboring particles and hard core repulsion at short distances. Another important ingredient is the tendency of active particle to align their velocities. Finally, the dynamics are affected by noise.

To make a concrete example of an active particle model used to describe wound healing, we describe here the model introduced by Sepúlveda et al. (2013). The equation of motion for each cell i is given by

$$\frac{d\mathbf{v}_i}{dt} = -\alpha \mathbf{v}_i + \sum_j \left[\frac{\beta}{N_i}(\mathbf{v}_j - \mathbf{v}_i) + \mathbf{f}_{ij} \right] + \sigma(\rho_i)\eta_i + \mathbf{F}_{\mathrm{rf}}(\mathbf{x}_i), \qquad (7.12)$$

where the sum is restricted to the nearest neighbors of i, α is a damping parameter, β is the velocity coupling strength and $\mathbf{f}_{ij}$ is the force between neighboring cells, taking into account short-range repulsion and long-range attraction. The noise is a random force $\sigma(\rho_i)\eta_i$, where η_i is an Ornstein–Uhlenbeck process with correlation time τ:

$$\tau \frac{d\eta_i}{dt} = -\eta_i + \xi_i, \qquad (7.13)$$

ξ_i is a delta-correlated white noise, independent for each cell $\langle \xi_i(t)\xi_j(t') \rangle = \delta_{ij}\delta(t - t')$. The amplitude of the noise term σ depends on the density of the neighboring particles ρ_i:

$$\sigma(\rho_i) = \sigma_0 + (\sigma_1 - \sigma_0)(1 - \rho_i/\rho_0). \tag{7.14}$$

The neighbors of each cell i are defined by first splitting the neighborhood of each cell into six equal sectors inside a given radius R and then finding the cell in each sector that is closer to i. In Equation 7.14 ρ_i and ρ_0 are the local and global particle densities, respectively. To model the invasion dynamics, Sepúlveda et al. (2013) introduce a free surface modeled by surface particles that are hindering cells in entering the empty space. The interaction between a surface particle and a cell is modeled with a repulsive potential. Prolonged contact between particles and cells leads to the damage of the latter, allowing cells to invade. Numerical simulations of the active particle model allows the experimentally observed dynamics in epithelial wound healing assay to be reproduced with great accuracy (see Figure 7.14).

Sepúlveda et al. (2013) also use the model to study leader cells, a subset of cells with characteristic phenotype such as a larger velocity. The concept of leader cell has been applied in several contexts such as wound healing, morphogenesis and cancer invasion (Khalil and Friedl, 2010). According to this idea, collective cell migration is driven by a subset of pathfinder or leader cells that explore the environment, select the best way and actively generate traction forces needed to move. While in individual cell migration all these tasks are performed by all the cells independently, in collective cell migration only a few cells would coordinate the process, interacting with the other cells through cell–cell junctions.

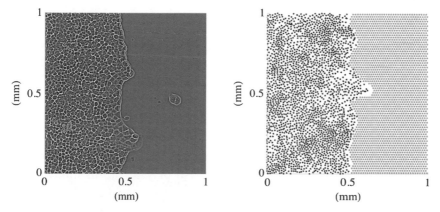

Figure 7.14 Comparison between experiments on epithelial wound healing and simulations of an active particle model. Image by Sepúlveda et al., 2013, CC-BY-4.0 licence.

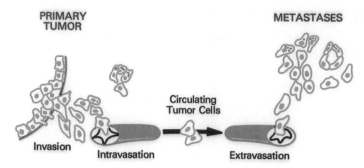

Figure 7.15 The main steps characterizing the metastatic process. Image by the National Cancer Institute, public domain.

7.6 Physics of Cancer Metastasis

Cancer metastasis involves a complex combination of many of the processes discussed in the present chapter. A primary tumor undergoes a series of steps before becoming metastatic (see Figure 7.15). It first becomes vascularized through angiogenesis and vasculogenic mimicry (see Chapter 4). Then cells from the primary tumors detach from the mass, penetrate a blood vessel or *intravasate* and circulate through the vascular system. The cells that are able to navigate successfully the blood stream are usually referred to as circulating tumor cells (CTC). Some of the cells attach to the blood vessel walls and exit or extravasate into a different tissue where they give rise to a secondary tumor. All these steps involve the physical and mechanical forces that are needed in order to migrate and flow through different environments (Wirtz et al., 2011). To reach a blood vessel, cancer cells must first overcome the physical barriers posed by the ECM, then considerably deform elastically to change their shape, squeeze in and out of the vessel, and then resist blood flow in order to adhere to the vessel walls.

Once cancer cells detach from the primary tumor, they face the complex physical barrier posed by the ECM. To this end, cancer cells release matrix metalloproteinases that dissolve structural components of the ECM. Cancer cells can also deform to pass through the porous fiber network, pulling on the fibers by forming integrin-based focal adhesion. Migration experiments in reconstructed collagen networks show the formation of pseudopodal protrusions, pulling in alternation at the front and back of the cell, thus enabling forward motion (Doyle et al., 2009; Wirtz et al., 2011). Cell mechanics also play a role in intravasation and extravasation: tumor cells undergo shape changes and cytoskeletal rearrangements to squeeze through endothelial cell junctions. As discussed in Section 6.2, cancer cells are usually softer than non-cancer cells, possibly enhancing their invasion capabilities. Wirtz et al. (2011) argue that intravasation occurs only for optimized

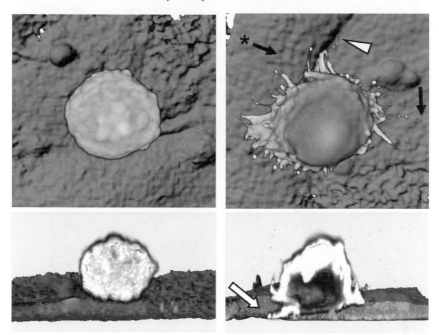

Figure 7.16 Confocal microscopy images of human mammary epithelial cells using microtentacles (black arrows) to intravasate, penetrating endothelial cell junctions (white arrowhead). Image by S. S. Martin, National Cancer Institute, public domain.

cell mechanical properties, but due to the broad heterogeneity of cancer cell populations allows the selection of cells with properties that are suitable for metastasis. In addition to deforming, cancer cells work actively to penetrate endothelial cell junctions by forming microtentacles (Whipple et al., 2010), as shown in Figure 7.16.

In the blood stream, CTCs should survive under dire conditions, resisting flow stresses, the immune response and collisions with other cells, so that only a small fraction of them can ultimately lead to metastasis. Physical parameters, such as fluid viscosity, and geometrical parameters, such as the vessel diameter, should play a fundamental role through this journey (Wirtz et al., 2011). Surviving CTCs should then extravasate by first attaching to the vessel wall. CTCs can stop when reaching sufficiently small vessels (Wirtz et al., 2011), but otherwise this process can occur with success when the adhesion strength overcomes for a sufficiently long time the flow-induced shear stress. The extravasation of CTCs is in most cases a complex dynamical process in which the cell collides with the walls, rolls over the vessel, arrests and finally manages to extravasate (Wirtz et al., 2011).

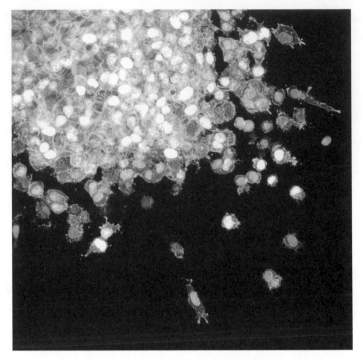

Figure 7.17 Individual lung cancer cells detach from a tumor spheroid and invade the surrounding environment (in this case a 3D collagen matrix). Image by S. Wilkinson and A. Marcus, National Cancer Institute, public domain.

The main conceptual framework to interpret the metastatic process is based on the idea that each metastasis arises from the clonal growth of a single tumor cell that has detached from tumor mass and migrated elsewhere (see Figure 7.17). This could be the result of an epithelial-mesenchymal transition (EMT) and the migration of a cancer stem cell. Recent genetic evidence indicates, however, that metastases can contain multiple clones (Gundem et al., 2015), supporting the notion that they arise from collective cell migration. This is in line with earlier studies detecting clusters of tumor cells in the bloodstream of cancer patients (Moore et al., 1960) and others that showed that clusters of cells could successfully seed metastasis (Liotta et al., 1976) (see Figure 7.18). More recently, Cheung et al. (2016) address two important unresolved questions: how do tumor cell clusters emerge from the primary tumor, and which molecular features identify cell clusters that ultimately give rise to metastasis? The first very important observation is that in the vast majority of epithelial tumors, cancer cells form cohesive clusters that collectively invade the surrounding stroma (Bronsert et al., 2014; Friedl et al., 2012). In breast cancer, collective invasion appears to be driven by a subset of leader cells expressing a high level of keratin 14 (K14) and other epithelial

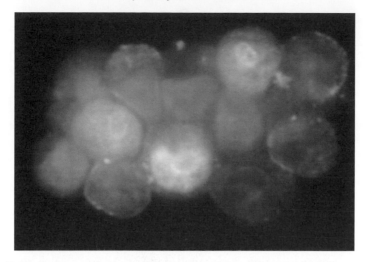

Figure 7.18 A cluster of circulating tumor cells from the blood of a breast cancer patient. Image by M. Yu, National Cancer Institute, public domain.

markers (Cheung et al., 2013). Cheung et al. (2016) show that K14 has a crucial role to drive cell invasion by inducing cell protrusions, while E-cadherin ensures cohesion within the cell cluster. Furthermore, it appears that K14 is required for collective but not for individual invasion. It is also possible that invasion crosses over from individual to collective cellular organization before intravasation in order to reach a survival advantage during metastasis.

8

Chromosome and Chromatin Dynamics in Cancer

Cancer typically displays alterations in the number of chromosomes, often the result of incorrect segregation during cell division, as we discuss in Section 8.1. Faithful chromosome segregation relies on the combined action of motor proteins and microtubules which must first align all the chromosomes on the cell central plate. This complex process can be described realistically by computational models which allow the effect of several biological factors on cell division to be tested, as we discuss in Section 8.2. In addition to chromosomal alterations, cancer cells often display other defects in nuclear architecture and chromatin organization, as we review in Section 8.3.

8.1 Chromosomal Instability

As discussed previously, cancer results from the accumulation of genetic mutations, but how these mutations are generated is debated. A high frequency of mutations, known as genetic instability, is believed to be a key property, if not a requirement, of most tumors (Lengauer et al., 1998). Genetic instability exists at the level of the nucleotides, resulting from their insertions, delations or substitutions, or at the level of the entire chromosome. Indeed, most cancers display an altered number of chromosomes, a state known as *aneuploidy*, and mis-segregated chromosomes at high rates, a condition known as *chromosomal instability* (CIN) (see Figure 8.1). Aneuploidy is simply detected by counting chromosome numbers, a task that can easily be achieved by several experimental techniques (Thompson et al., 2010). Detection of CIN requires instead the measurement of chromosome mis-segregation rates, which involves counting the number of chromosomes at different times in clonal populations.

CIN is a characteristic feature of human solid tumors and of many hematological malignancies (Boveri, 1903), a principal contributor to genetic heterogeneity in cancer (Burrell et al., 2013a) and an important determinant of clinical prognosis

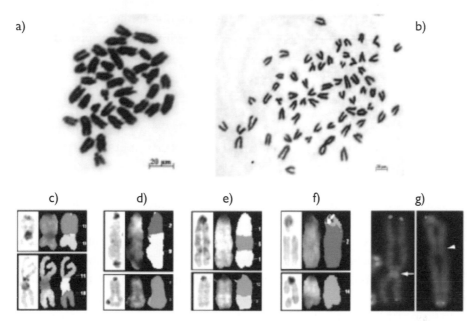

Figure 8.1 Gross chromosomal instabilities in Hltf-/-/Apcmin/+ colon tumors as revealed by cytogenetic analysis. a) Representative metaphase spread of Apcmin/+ colon tumor cells as revealed by Giemsa staining, showing a near-diploid karyotype without obvious chromosomal abnormalities in these tumor cells. b) Giemsa staining of metaphase spreads prepared from Hltf -/-/Apcmin/+ colon tumor cells, which displayed aneuploid and gross chromosomal instability. c-f) Some commonly detected chromosomal abnormalities in Hltf -/-/Apcmin/+ colon tumor cells. c) Robertsonian fusions; d) Dicentric chromosomes; e) Unbalanced chromosomal translocations; f) Chromosomal breaks. g) Telomere dysfunction as revealed by Q-FISH. A white arrow indicates chromosomal fusion without telomeres at the fusion site, and a white arrowhead points to lack of telomere signals at the fusion site of a dicentric chromosome ring. Image by Sandhu et al., 2012, CC-BY-2.0 licence.

and therapeutic resistance (Lee et al., 2011; Bakhoum and Compton, 2012). While the link between aneuploidy and cancer was recognized already over a century ago (Boveri, 1903), general understanding of the mechanisms leading to CIN, as well as appreciation of its consequences on cellular viability and tumor evolution, has grown considerably over the past two decades (Bakhoum and Compton, 2012; Thompson et al., 2010). Multiple mechanisms have been shown to lead to CIN, including defects in the spindle assembly checkpoint (Cahill et al., 1998), the regulation of microtubule attachments to chromosomes at kinetochores (Bakhoum et al., 2009a,b), centrosome duplication (Silk et al., 2013), sister chromatid cohesion (Zhang et al., 2008a), telomere maintenance (Davoli et al., 2010) and pre-mitotic replication stress (Burrell et al., 2013b).

In broad terms, CIN can be caused by misregulation in the centrosome, where MTs are nucleated, or at the level of the kinetochore, where microtubule tips attach. For instance, the centrosomal protein 4.1-associated protein (CPAP), which belongs to the microcephalin (MCPH) family (Bond et al., 2005), localizes to the site of the growing centriole (Lawo et al., 2012) and inhibits microtubule nucleation (Hung et al., 2004). The rate of microtubule nucleation has recently been enhanced by overexpression of the oncogene AURKA or by loss of the tumor suppressor gene CHK2, leading in both cases to CIN (Ertych et al., 2014). Another example is provided by mitotic centromere-associated kinase or kinesin family member 2C (MCAK/Kif2C) that is localized at the centromeres, kinetochores and spindle poles, and functions as a key regulator of mitotic spindle assembly and dynamics (Hunter et al., 2003; Zhu et al., 2005). Depletion or down-regulation of MCAK/Kif2C revealed chromosomal congression and segregation defects due to improper kinetochore attachments (Bakhoum et al., 2009a; Stout et al., 2006). In contrast, overexpression of MCAK/Kif2C enhances microtubule depolymerization, resulting in their detachment from centromeres (Ganguly et al., 2011). Higher expression of MCAK level has been found in gastric cancer tissue (Nakamura et al., 2007), colorectal and other epithelial cancers (Gnjatic et al., 2010) and breast cancer (Shimo et al., 2008).

As discussed in Section 1.7, chromosome segregation is achieved by the coordinated action of microtubules and molecular motors that assemble the chromosomes at the central plate of the cell (see Figure 8.2). In order to correctly segregate, the two sister kinetochores of each chromosome should be correctly attached to the two

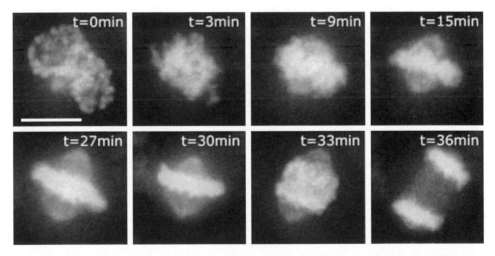

Figure 8.2 The process of chromosome congression during cell division. Chromosomes are initially scattered and should first be assembled at the center of the cell. Image by Bancroft et al., 2015, CC-BY-3.0 licence.

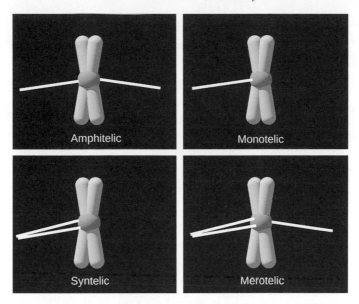

Figure 8.3 Microtubule-kinetochore attachments. Correctly (amphitelic) attached sister kinetochores orient toward opposite spindle poles. Faulty attachments include cases where only a single kinetochore is attached to a spindle pole (monotelic attachment), both sister kinetochores are attached to the same pole (syntelic attachment) and one sister kinetochore is attached to two poles (merotelic attachment).

poles (amphitelic attachment). When this spindle assembly checkpoint is passed, sister chromatid cohesion is disrupted and sister chromatids can successfully segregate. Errors can occur due to faulty attachments (see Figure 8.3): monotelic attachments when only one sister kinetochore is attached, syntelic attachments when both sister kinetochores are attached to a single pole and merotelic attachments when one sister kinetochore is attached to two poles. Occasional merotelic attachments are usually corrected prior to anaphase, but are not sensed by the spindle assembly checkpoint. Hence, if correction mechanisms do not work properly, both sister chromatids can be transported to the same pole, yielding chromosome mis-segregation.

A central role in faulty attachment correction is played by the centromere-localized Aurora B kinase. The prevailing view is that Aurora B activity is modulated by the mechanical tension between sister kinetochores (Liu et al., 2009; Lampson and Cheeseman, 2011). Aurora B kinase generates a phosphorylation gradient suppressing the activity of MCAK and recruiting Kif2b, increasing the detachment rate of microtubules (Fuller et al., 2008). When chromosomes biorient, the sister kinetochores are under tension and the phosphorylation gradient cannot reach the target protein, leading to a stabilization of the microtubule-kinetochore attachments. Tumor cells with CIN have been shown to display a higher stability in microtubule-kinetochore attachments, suggesting that error-correcting mechanisms do not work properly (Bakhoum et al., 2009a).

8.2 Theoretical and Computational Models of Chromosome Segregation

Chromosome congression occurs in a rapidly fluctuating environment since the mitotic spindle is constantly changing due to random microtubule polymerization and depolymerization events. This process, known as dynamic instability, is thought to provide a simple mechanism for microtubules to search-and-capture all the chromosomes scattered throughout the cell after nuclear envelope breakdown (NEB) (Kirschner and Mitchison, 1986). Once chromosomes are captured, they are transported to the central plate by molecular motors that use microtubules as tracks. The main motor proteins implicated in this process are kinetochore dynein, which move towards the spindle pole (i.e., the microtubule minus end) (Rieder and Alexander, 1990; Li et al., 2007; Yang et al., 2007; Vorozhko et al., 2008), and centromere protein E (CENP-E or kinesin-7) (Kapoor et al., 2006; Cai et al., 2009; Barisic et al., 2014) and polar ejection forces (PEFs) (Rieder ct al., 1986), both moving away from the pole (i.e., they are directed towards the microtubule plus end). PEFs mainly originate from kinesin-10 (Kid) and are antagonized by kinesin-4 (Kif4A) motors (Stumpff et al., 2012), sitting on chromosome arms (Wandke et al., 2012). While PEFs are not necessary for chromosome congression, they are vital for cell division (Barisic et al., 2014) since they orient chromosome arms (Wandke et al., 2012), indirectly stabilize end-on attached microtubules (Cane et al., 2013) and are even able to align chromosomes in the absence of kinetochores (Cai et al., 2009). Recent experimental results show that chromosome transport is first driven towards the poles by dynein and later towards the center of the cell by CENP-E and PEF (Barisic et al., 2014) (see Figure 8.4).

A quantitative understanding of chromosome congression has been the goal of intense theoretical research focusing on the mechanisms for chromosome search-and-capture (Holy and Leibler, 1995; Wollman et al., 2005; Paul et al., 2009), motor driven dynamics (Joglekar and Hunt, 2002; Civelekoglu-Scholey et al., 2006; Gardner et al., 2005; Chacón and Gardner, 2013; Glunčić et al., 2015) and attachments with microtubules (Hill, 1985; Bertalan et al., 2014). A mathematical study of search-and-capture was performed by Holy and Leibler (1995), who computed the rate for a single microtubule to find a chromosome by randomly exploring a spherical region around the pole. Later, however, Wollman et al. (2005) showed numerically that a few hundred microtubules would take about an hour to search and capture a chromosome, instead of a few minutes as observed

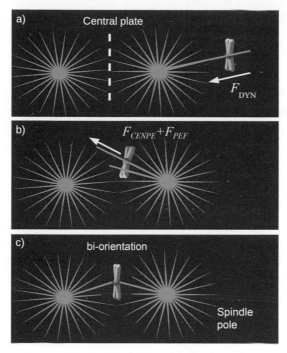

Figure 8.4 Schematic of the dynamics of a single chromosome. a) Peripheral chromosomes, not lying between the spindle poles, are driven to the nearest pole by dynein. b) Chromosomes are driven from the pole to the central plate by the combined action of CENP-E and PEF. c) At the central plate, chromosomes attached to both poles are called bi-oriented. Image by Bertalan et al., 2015, CC-BY-4.0 licence.

experimentally. It was therefore argued that microtubules should be chemically biased towards the chromosomes (Wollman et al., 2005). An alternative mechanism proposed to resolve this discrepancy is the nucleation of microtubules directly from kinetochores (O'Connell and Khodjakov, 2007), which was incorporated in a computational model treating chromosomal movement as random fluctuations in three dimensions (Paul et al., 2009).

Describing motor-driven chromosome dynamics and microtubule attachment (Hill, 1985; Bertalan et al., 2014) has also been the object of several computational studies mainly focusing on chromosome oscillations (Joglekar and Hunt, 2002; Civelekoglu-Scholey et al., 2006). These one-dimensional models do not account for congression, because they do not consider peripheral chromosomes, not lying between the spindle poles at NEB, which are, however, experimentally observed in mammalian cells (Barisic et al., 2014). Three-dimensional numerical models have been extensively introduced to study cell division in yeast (Gardner et al., 2005; Chacón and Gardner, 2013; Glunčić et al., 2015), but in that case motor proteins are not essential for congression and there is no NEB. It is therefore not clear to what extent these models can be applied to mammalian cells.

The concerted action of deterministic transport by molecular motors and stochastic microtubule growth and shrinkage kinetics to drive successful chromosome congression can be successfully captured by a recent three-dimensional computational model (Bertalan et al., 2015). The model describes accurately the processes of stochastic search-and-capture by microtubules and deterministic motor-driven transport, reproducing accurately experimental observations obtained when individual motor proteins were knocked down (Antonio et al., 2000; Kapoor et al., 2006; Putkey et al., 2002; Silk et al., 2013; Barisic et al., 2014) (see Figure 8.5), demonstrating the crucial role played by the number of microtubules to achieve successful chromosome congression and bi-orientation. Increasing the number of microtubules improves bi-orientation but slows down congression. Conversely, when the number of MTs is too low, congression probability is increased but bi-orientation is impaired (see Figure 8.6). The numerical value of the optimal number of microtubules is estimated to be around 10^4, in agreement with experimental measurements (McIntosh et al., 1975) but much larger than the number of microtubules employed in previous computational studies (Magidson et al., 2011; Wollman et al., 2005; Paul et al., 2009).

The model shows that coherent and robust chromosome congression can only happen if the total number of microtubules is neither too small nor too large. The results allow for a coherent interpretation of many biological factors already associated in the past with CIN (Bertalan et al., 2015). For instance, the centrosomal protein 4.1-associated protein (CPAP), belonging to the microcephalin (MCPH) family (Bond et al., 2005), is known to inhibit MT nucleation (Hung et al., 2004) and its overexpression leads to abnormal cell division (Kohlmaier et al., 2009; Schmidt et al., 2009). In the model, CPAP overexpression is simulated by inhibiting microtubule nucleation, which can decrease the number of microtubules below the optimal value, leading to mitotic errors. Similarly, the model explains the role of mitotic centromere-associated kinase or kinesin family member 2C (MCAK/Kif2C). Both depletion (Stout et al., 2006; Bakhoum et al., 2009a) and overexpression (Ganguly et al., 2011; Tanenbaum et al., 2011) of MCAK lead to cell division errors. MCAK/kif2C is localized at microtubule plus ends (Domnitz et al., 2012) and functions as a key regulator of mitotic spindle assembly and dynamics (Hunter et al., 2003; Zhu et al., 2005) by controlling microtubule length (Domnitz et al., 2012). Higher expression of MCAK level has been found in gastric cancer tissue (Nakamura et al., 2007), colorectal and other epithelial cancers (Gnjatic et al., 2010) and breast cancer (Shimo et al., 2008). MCAK overexpression can be modeled by increasing the

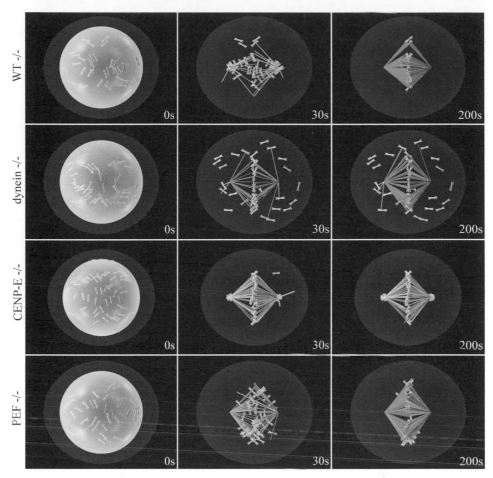

Figure 8.5 Time-lapse snapshots of the simulated congression process when motors are suppressed. Not all microtubules are shown, only those that serve as rails for kinetochore motor-proteins and end-on attached MTs. The nuclear envelope is shown for reference in each of the first panels as a white sphere. The cortex is represented in dark grey. The wild type (WT) case, in which all motor proteins are active, is shown for comparison in row 1. When dynein is suppressed (row 2), PEFs push peripheral chromosomes to the cortex. However, when all chromosomes start between the poles, congression takes place normally. When CENP-E is depleted (row 3), peripheral chromosomes or other chromosomes that are transported to the poles get trapped there. Depleting PEFs (row 4) delays congression significantly and destabilizes the coherence of the central plate. It makes no difference whether all chromosomes start between poles or there are peripheral chromosomes. Image by Bertalan et al., 2015, CC-BY-4.0 licence.

rate of microtubule depolymerization, which reduces their length and number to a level in which bi-orientation is not possible. Finally, the model explains recent results linking CIN to the overexpression of AURKA or the loss of CHK2, both enhancing microtubule assembly rate (Ertych et al., 2014). Increasing

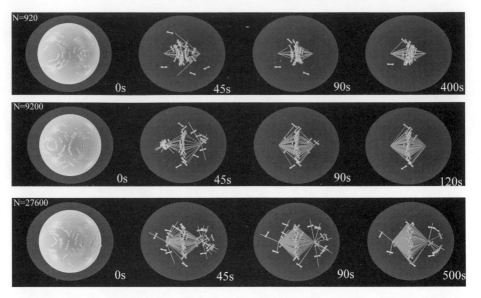

Figure 8.6 Time-lapse snapshots of the simulated congression process for different values of the number of microtubules per kinetochore. Congression fails if this number is too small or too large. Chromosomes are shown as having chromatid arms for viewing purposes, while the kinetochores are shown as spheres. Not all microtubules are shown; only those that serve as rails for kinetochore motor-proteins and end-on attached MTs. The nuclear envelope is shown for reference in each of the first panels as a white sphere. The cortex is represented in dark grey. From Bertalan et al. (2015) CC-BY-4.0 licence.

microtubule velocity effectively reduces the amount of tubulin units available for microtubule nucleation, thus decreasing the number of microtubules and impairing bi-orientation.

8.3 Nuclear Alterations in Cancer

In addition to CIN, cancer cells are often characterized by an altered nuclear architecture (Zink et al., 2004). Cancer diagnosis has relied for decades on the observation of the nuclear morphology, which in cancer cells displays size and shape variations as well as changes in chromatin structure (Zink et al., 2004).

As briefly discussed in Chapter 1, the cell nucleus is enveloped by a lamin-based scaffold surrounded by a double lipid bilayer on the cytoplasmic side and bound to chromatin on the inner side. The inner side of the nuclear envelope binds to chromatin fibers, and in particular to heterochromatin domains, which in cancer cells often appear to be altered either because of their loss or because of their displacements. An example is reported in Figure 8.7, showing that cells in tissues displaying precancerous adenoma in the colon show a significant decrease in

(a) (b)

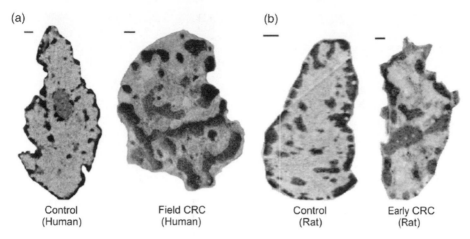

| Control
(Human) | Field CRC
(Human) | Control
(Rat) | Early CRC
(Rat) |

Figure 8.7 Transmission electron micrograph of the nuclei from human and rat tissues, either normal or with precancerous adenoma. Dark regions correspond to heterochromatin. Scale bar is a) 500nm, b) 250nm. Image by Cherkezyan et al., 2014, CC-BY-4.0 licence.

the heterochromatin at the periphery and a simultaneous increase in the nuclear interior (Cherkezyan et al., 2014). Alterations in chromatin structure are known to affect gene expression (Zink et al., 2004). For instance, Chalut et al. (2012) show correlations between the expression of the pluripotent marker Nanog, the condensation state of chromatin and the mechanical stiffness of the nucleus in embryonic stem cells. One possible (but still debated) mechanism could be the activation of mechanosensitive genes through cytoskeleton–nuclear couplings (Denais and Lammerding, 2014). On the other hand, it is clear that chromatin organization affects the mechanical properties of the nucleus and the migration properties of the cell.

Normal cells typically show regular ellipsoidal nuclear shapes which in cancer cells become more irregular and suggestive of a more flexible structure. This observation could have implications for metastasis (Denais and Lammerding, 2014). As discussed in Section 7.6, cancer cells need to overcome several geometrical and physical barriers in order to navigate through the ECM and initiate the metastatic process. Typical pore sizes in the ECM vary considerably, from large ones spanning tens of microns to smaller ones of micron and submicrometer scale. While the cytoplasm and the plasma membrane can be remodeled to squeeze through narrow pores, the nucleus is much stiffer and is thus a limiting factor for cell migration (Friedl et al., 2011; McGregor et al., 2016). An increased nuclear flexibility improves the migration capability of the cells and therefore their metastatic potential. The mechanical deformation of the nucleus when cells traverse a narrow barrier is quite intriguing (Figure 8.8): first the nuclear membrane

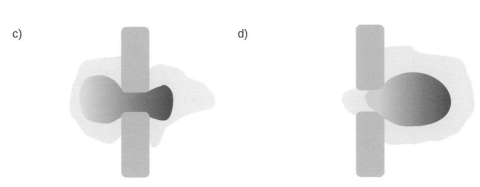

Figure 8.8 Nuclear deformation as a cell squeezes through a narrow pore. a) The nucleus is compressed against the pore walls, b) locally deforms, c) changes conformation and d) passes through the obstacle. See Friedl et al. (2011) for more details.

is compressed against the pore walls, leading to local deformation and followed by a dramatic remodeling of the nucleus, which assumes an hourglass shape before finally squeezing through the pore (Friedl et al., 2011). Recent experiments show that cancer cells can deform to the point of breaking the nuclear envelope to pass through a very narrow pore (Denais et al., 2016; Raab et al., 2016). The nuclear envelope is rapidly self-repaired thanks to an ESCRT (endosomal sorting complexes required for transport) based mechanism (Denais et al., 2016; Raab et al., 2016).

9

Control of Tumor Growth by the Immune System

The immune system is the primary defense of the organism against diseases, but cancer is often able to evade its response. Many advanced therapeutic strategies nowadays aim to redirect the immune response against the tumor. In Section 9.1, we introduce the immune system, discussing innate and adaptive immunity. The ability of cancer cells to escape the immune response is discussed in Section 9.2. The communication between cells during the immune response is coordinated by cytokines, as we discuss in 9.3. The organism responds to external pathogens by a process of inflammation that we review in Section 9.4, where we also discuss its relevance for tumors. Finally, in Section 9.5, we illustrate simple mathematical models for the interactions between cancer cells and the immune system.

9.1 The Immune System

The immune system plays the crucial role of defending organisms against not only pathogens but also tumor cells. In mammals, immuno-surveillance is a combination of innate and adaptive immunity (Figure 9.1). The main feature of innate immunity is to activate an immediate response making use of pre-existing mechanisms: mechanical barriers, enzymes, complementary systems and cells, such as neutrophils that can phagocyte pathogens, or like the natural killer (NK) that can kill them. Adaptive immunity is provided by lymphocytes B and T that require first to be activated by pathogens and only then proliferate, finally inducing terminal factors needed to eliminate the pathogens. The whole process of adaptive immunity requires more time than innate immunity, but once the system is immunized, the response is much faster and stronger thanks to the release of memory cells (Figure 9.2). In the adaptive immunity, the antigen should be presented by Major Histocompatibility Complex (MHC) expressed or on the plasma membrane of all the nucleated cells (MHC type I) or on the surface of the antigen presenting cells (MHCII) at the lymphocyte T that recognize the complex MHC-peptide through

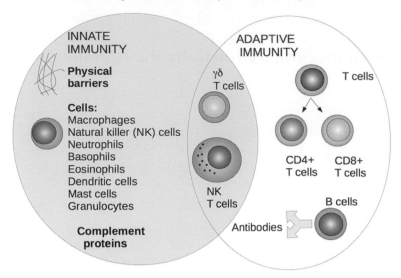

Figure 9.1 Schematic of the actors playing a role in innate and adaptive immunity. See Dranoff (2004) for more details.

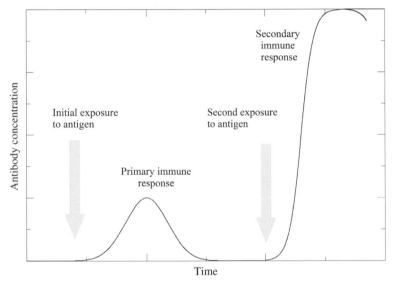

Figure 9.2 Time evolution of adaptive immune response. The initial response is slow, but once the system is immunized the response is faster and stronger.

the T Cell Receptor (TCR). In contrast, B cells recognize directly the antigen through the antibody expressed on the plasma membrane (B Cell Receptor, BCR).

Pathogens are eliminated by an interplay between innate and adaptive immunity, which act cooperatively. The communication between the two immunities, and in general the coordination of all the immune responses, is due to a complex network of signals mediated by cytokines (Dranoff, 2004). These are small proteins that are expressed by the immune cells in a time-dependent manner. Cytokines are at the same time redundant, since they act on many targets, and pleiotropic in that many of them share the same target. All these biological features endow the immune system with a great robustness. The possibility for the immune cells, once activated by the antigen, to reach the target is co-ordinated by the chemokines. These small proteins are a subclass of cytokines involved in chemotaxis. In recent years, an enormous body of literature has been devoted to the role of each of the cytokines, investigating the cells that release them and on which kinds of cells they act. The picture is quite complex, and it is now important to combine all these factors together in order to understand what happens when there is an imbalance in one or many of the cytokines: an issue that is crucial for many pathological conditions, from cancer to autoimmunity.

Autoimmunity, the chance of activating the immune system by self-antigens, is avoided by two mechanisms. The first is central and is based on death through apoptosis of auto-reactive clones at the level of bone marrow for B cells and of the thymus for T cells. The other mechanism is present at the periphery and is a consequence of the fact that autoreactive clones are not responsive, due to a biological mechanism called *anergy*. Autoimmunity is the breakdown of tolerance to the self that leads to the activation of the adaptive immunity against self-antigens. For instance, adaptive immunity is activated by antigens present in transplanted tissues, making it difficult to share organs between individuals of the same species. In principle, immuno-surveillance also exists against cancer cells, as we discuss below.

9.2 Immunogenicity of Cancer Cells

Immunogenicity is conventionally defined as the capability to induce adaptive immune responses. The same concept holds for tumors, where one speaks of tumor immunogenicity as the ability to induce an immune response preventing tumor growth. Tumors can in principle be rejected by their host, but they can become tolerogenic, thus evading the immune response through different biological mechanisms. When this happens adaptive immunity cannot eradicate them, even if it is able to recognise them. In general, tumors are considered highly immunogenic when they grow well in naive syngeneic mice and when there is evidence of intense immuno-response in bioptic samples.

Tumors generally carry antigens that can be targeted by T cells. Some tumors, however, express more antigens than others. This is sometimes attributed to differences in the expression of human leukocyte antigens (HLA) and to DNA methylation of the cancer-germline genes that regulate them, or to the loss of DNA mismatch repair leading to more antigenic mutations, as in colorectal carcinoma (Kloor et al., 2010). Immune interventions are sometimes effective in melanoma, probably because of the expression of melanocyte-specific antigens. In general, however, the cancer stroma secretes immunosuppressive factors and recruits suppressive myeloid and lymphoid immune cells, creating an environment where the immune response is hindered. Tumor cells also produce immunosuppressive cytokines such as TGF-beta or NK decoys. Finally, immune cells are prevented to access the tumor by molecular barriers produced by endothelial cells, a dense ECM and high IFP due to insufficient lymphatic drainage.

Many immunomodulatory mechanisms operate early in cancer development but resistant variants that evade immunosuppression can emerge, leading to downregulation of antigen processing. For immunotherapies to work, it is therefore imperative to act early, before the appearance of a population of immune resistant or immunosuppressive tumour cells which would lead to decreased tumour immunogenicity. Current immunotherapy is based on passive immunization through adoptive T cell transfer or active immunization through vaccination. Furthermore, one can block co-inhibitory signalling of tumor-specific T cells and combine these interventions with each other or with other classical anticancer therapies. It is still impossible, however, to understand precisely the reason why these interventions often fail. There are still many problems arising from the immunomodulatory mechanisms operating in tumors. We still do not know enough about the relevant mechanisms for each type of human cancer, especially because several mechanisms might operate simultaneously. We should also find reliable biomarkers to detect their activity and develop non-toxic methods for blocking them.

The most common vaccination protocols are designed to stimulate an immune response against antigens expressed by cancer cells (see Figure 9.3). Recent studies have shown that it is also possible to induce anti-tumor immunity using vaccination against products expressed by the typical constituents of the tumor environment, such as endothelial cells, macrophages or fibroblasts (Nair et al., 2003; Niethammer et al., 2002). The first vaccine for the treatment of prostate cancer and the first immunomodulating agent against melanoma were recently introduced (Kantoff et al., 2010), and there is thus hope that these two innovative clinical attempts could lead to the development of additional vaccines and immunomodulating agents to enhance the immune recognition of tumor cells expressing immunogenic antigens. The success of future tumor immunotherapy strategies will probably depend on the identification of novel and more effective immunogenic antigens to target tumors.

Although the immunogenicity of an antigen is crucial, it is equally important to target antigens that are biologically important to tumor progression. Therefore, we need to better understand how potentially immunogenic antigens are linked to the

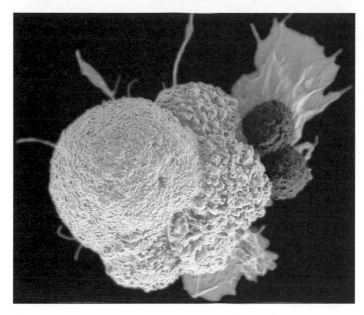

Figure 9.3 Immune system response to cancer cells. The image shows a scanning electron micrograph of an oral squamous cancer cell attacked by two cytotoxic T cells (dark). Image by R. T. Serda, National Cancer Institute, Duncan Comprehensive Cancer Center at Baylor College of Medicine, public domain.

biological properties of cancer. Therapeutic success will also depend on developing the best antigen delivery systems. Issues concerning the optimal vaccine platform (peptide, DNA, protein, viral, bacterial and so on) and the number of antigens that should be simultaneously targeted still need to be addressed. Finally, we still need to elucidate the entire network of immune signaling pathways regulating immune responses in the tumor microenvironment. These regulatory mechanisms are probably specific for each type of cancer, which has its own unique set of genetic, epigenetic and inflammatory changes evolving with the advancement of the disease. Understanding these networks will provide new targets for modulation and, ultimately, will result in improved combinatorial immunotherapies for a successful cancer treatment. The discovery of CSC subpopulations in tumors could open new perspectives also for immunotherapy. In melanoma, CSCs express the same antigens as the other cancer cells (Monzani et al., 2007). Therefore, from the immunological point of view, CSCs are not peculiar. It remains to be seen if this feature is general also for other tumors.

9.3 Cytokines and the Regulation of the Immune System

The immune response is coordinated by cytokines, a large family of small signaling proteins which stimulate the communication between cells. In pathological processes such as inflammation, infection or trauma, cytokines are unbalanced, thus playing a critical role. The presence of immune stimuli leads to *de novo* secretion of cytokines acting for a limited time and in a well-defined location, usually at very low concentrations. The only exceptions are the cytokines acting as an endocrine hormone, whose action is distant from the point of release.

Cytokine receptors are usually subdivided into seven classes:

1. Type I cytokine receptors (i.e., IL1, GM-CSF etc);
2. Type II cytokine receptors (IL2, interferons);
3. Immunoglobulin superfamily (Il1, Il18 etc.);
4. Tumor Necrosis Factor (TNF) Receptor Family (CD27, CD120 etc.);
5. Chemokine receptors (Il8, CCR1, CXCR4 etc.);
6. TGFβ receptors (TGFβ Receptor 1, TGFβ Receptor 2).

These classifications reflect some common properties within each class: Type I and Type II receptors show certain conserved motifs in their extracellular amino-acid domain. Immunoglobulin superfamily receptors share structure with immunoglobulins, adhesion molecules, etc.; TNF receptor family receptors share cysteine-rich common extracellular binding domains; chemokine receptors have seven transmembrane helices.

Cytokines act either in an autocrine manner (on the same cells that produce the cytokine) or in a paracrine manner (on neighboring cells) if those cells express the receptor. Moreover, the action of cytokines is pleiotropic, redundant, synergic and antagonistic: pleiotropic because of their ability to induce varied responses in target cells, such as proliferation, differentiation or migration; redundant since several cytokines induce the same response; synergic due to the ability of different cytokines to cooperate in inducing certain cell responses. Finally, certain cytokines are able to antagonise the activity of other cytokines: the robustness of the immune system results from the combination of all these properties. Chemokines are a subclass of cytokines whose role is to induce directed chemotaxis in nearby responsive cells, as already discussed in Section 7.2. They are very small with a molecular mass between 8 and 10kDa (see Figure 9.4) and are divided into two families (C chemokines or CX3C chemokines), based upon the contiguity or separation of cysteine residues (CXC). The majority of chemokines guide the migration of cells such as homing of lymphocytes and many are involved in inflammation.

9.4 Inflammation in Tumors

Inflammation is a complex biological response of body tissues to harmful stimuli, such as pathogens, heat or toxic substances. The role of inflammation is to eliminate the initial cause of cell injury, clear out the necrotic tissue and finally

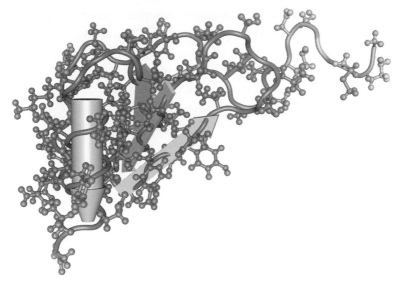

Figure 9.4 Structure of the monomeric variant of the chemokine MIP-1β (Macrophage Inflammatory Protein). Image by N. Dilmen, CC-BY-SA-3.0 license.

restore the damaged tissues. There are two kinds of inflammation: acute and chronic. Acute inflammation can become chronic when the process persists for a long time. Acute inflammation involves a series of biochemical events occurring in the local vascular system, the recruitment of leukocytes, mainly neutrophyls and monocytes, and the triggering of several cellular biochemical cascade systems such as the complement, the kinin, the coagulation and fibrinolysis systems. Leukocytes extravasate from the vessels through active movements across endothelial cells, involving adhesion molecules and chemokines as chemoattractants. The cell adhesion molecules involved in a time dependent manner during inflammation are selectins and integrins (see Section 7.3 for more details on their structure and function).

It is quite clear that inflammation sustains tumor cells, by activating stroma and DNA-damage-promoting agents. While the connection between cancer and inflammation is widely accepted, the molecular and cellular mechanisms mediating this relationship are still unclear. The involvement of inflammation in tumor development can act at different levels, such as the release of cytokines which then help tumor cells or its contribution to immunoregulation. In this connection, tumor-associated macrophages derived from monocytes are recruited by monocyte chemotactic protein (MCP) chemokines and play a dual role in the tumor: on one side they kill tumor cells following activation by IL2, interferons and Il12, and on the other side they produce potent angiogenic and lymphoangiogenic factors, cytokines and proteases, thus enhancing tumor progression (Schoppmann et al., 2002). From the therapeutic point of view, the inhibition of inflammation represents a promising target. Long treatment with aspirin and nonsteroidal anti-inflammatory drugs show a drastic reduction of colon cancer of about 50 percent (Baron and Sandler, 2000). The literature reports multiple drugs that use as target pro-inflammatory cytokines such as TNAα, decreasing the level of expression of selectins (Shanahan and St Clair, 2002). Furthermore, the modulation of the environment and the relationship between food, inflammation and cancer is promising to become the object of intense investigation in the future. In this respect, there is a strong connection between chronic inflammation and malignant disease in colon carcinoma arising in individuals with inflammatory bowel disease such as Crohn's disease. The idea is that an infection caused by bacteria or a virus leads to a normal inflammatory response that is not resolved and becomes chronic.

9.5 Models for the Interaction Between Immunity and Tumors

The complexity of the immune system, with its many actors and the intricate stochastic kinetics, has fascinated physicists and mathematicians for a long time. A detailed early review of physics-based modeling efforts for the immune systems is reported in Perelson and Weisbuch (1997), where the authors discuss various kinetic network models and cellular automata. Perelson and Weisbuch (1997) conclude their paper expecting that key contributions to immunology coming from statistical physics should emphasize collective emergent properties. For instance, immune network models could help us understand how coordinated interactions among individual immune systems cells can generate a global immune response.

Several mathematical models of the interactions between the immune system and cancer have been studied in the past (Kirschner and Panetta, 1998; Robertson-Tessi et al., 2012; de Pillis et al., 2005; Mallet and De Pillis, 2006). These include, for instance, kinetic models of the T-cells' response to tumors (Kirschner and Panetta, 1998; Robertson-Tessi et al., 2012) or hybrid models of tumor-immune system interactions involving a cellular automaton coupled with discretized partial differential equations (Mallet and De Pillis, 2006).

To illustrate with a simple example of how these models are built we consider the model introduced by Kirschner and Panetta (1998) who considers activated immune cells, such as T-cells or natural killer cells, with concentration $E(t)$, tumor cells with concentration $T(t)$ and Interleukin-2, a cytokine responsible for lymphocyte activation, with concentration $I_L(t)$. The kinetic equations describing the population of these species are as follows:

$$\frac{dE}{dt} = cT - \mu_2 E + \frac{p_1 E I_L}{g_1 + I_L} + s_1 \tag{9.1}$$

$$\frac{dT}{dt} = r_2(T)T - \frac{aET}{g_2 + T} \tag{9.2}$$

$$\frac{dI_L}{dt} = \frac{p_2 ET}{g_3 + T} + s_1 - \mu_3 I_L + s_2. \tag{9.3}$$

Equation 9.1 describes the kinetics of effector cells that are triggered by the presence of tumor cells and stimulated by cytokines. The kinetics of tumor cells (Equation 9.2) contains a proliferation term and a loss term dependent on the concentration of effector cells. Finally, Equation 9.3 describes the kinetics of the cytokine that is stimulated by the presence of tumor cells. This relatively simple model already has several parameters that are difficult to estimate. Despite this, the model can be used to compute phase diagrams, allowing the parameter regions associated with distinct dynamical behaviors such as growth, decay or oscillations to be determined.

10

Pharmacological Approaches: Old and New

Traditional approaches to treat cancer rely on surgery, radiotherapy and chemotherapy. There is a current effort to improve the efficacy of drug delivery by using nanotechnology, encapsulating drugs inside nanoparticles. We discuss these methods in Section 10.1. Evidence of the relationship between cancer and nutrition has been accumulating over the years. We discuss this point in Section 10.2.

10.1 The Traditional Approaches and Nanomedicine

Current cancer treatments consist of surgery, radiotherapy and chemotherapy. Removal of the bulk of the tumor by surgery still remains the most effective treatment for cancer. Evidence exists, however, that surgery may induce an acceleration of tumor and metastatic growth due to the inflammatory response associated with wound healing (Coffey et al., 2003). Radio- and chemotherapy are often used to eliminate possible peripheral cells that are not completely eradicated by surgery and to control cancer growth activated by wound healing. Some tumors such as melanoma are radio-resistant, and therefore chemotherapy is the only possible alternative strategy apart from surgery.

Chemotherapy is not always effective due to problems in drug delivery, drug resistance and toxicity for the patients. Resistance to chemotherapy can have various causes in which a big role could be played by the large cell-to-cell variability inside a tumor, even within the same patient (Kessler et al., 2014). The current strategy is to use a combination chemotherapy, referring to the simultaneous administration of two or multiple therapeutic agents (Greco and Vicent, 2009; Dai and Tan, 2015; Xu et al., 2015; Pacardo et al., 2015). The main idea is that choosing an appropriate drug combination can help prevent cancer drug resistance and improve target selection and therapeutic action. An important limitation of combination chemotherapy stems from the different kinetics associated with each drug, making it difficult to obtain a simultaneous combined action. Furthermore, it is important to remark that the systemic toxicity of combination chemotherapy might be significantly enhanced due to the sum of side effects of separated drugs, which could enormously limit the effectiveness of combination therapy in the clinic.

Nanotechnology represents a promising strategy to improve drug delivery, allowing for easy drug administration, improved accumulation at the tumor site and at the same time minimization of side effects for the patient (Langer, 1998; Saltzman and Olbricht, 2002; Peer et al., 2007; LaVan et al., 2003; Farokhzad and Langer, 2009). Nanoparticles can be used as drug carriers, providing many advantages over the direct administration of free drugs (see Figure 10.1). In particular, they allow for the encapsulation of drug combinations into the same nanocarrier, thus reducing typical problems of combination therapy due to non-uniformity in pharmacokinetics (see Figure 10.2). Furthermore, nanocarriers can protect a drug from quick clearance by evading the reticuloendothelial system, allow for prolonged circulation times, increase the availability of drugs at the targeted site and enable transport through biological barriers. In this respect, the possibility to deliver drugs across the blood/brain barrier is of fundamental importance, since treating brain metastases is still an open issue. A careful nanoparticle design can help overcome barriers created by the tumor microenvironment and result in a better treatment. Indeed, drug delivery improved enormously thanks to the design of more versatile materials such as synthetic polymers, lipids and biomacromolecule scaffolds (Sarikaya et al., 2003; Allen and Cullis, 2004; Wagner et al., 2006; Shi et al., 2010; Smith et al., 2013). Nanoparticles selectively accumulate in tumors due to the increased permeability of tumor blood vessels facilitating the entrance in the tumor interstitial space. Furthermore, the dysfunction of the lymphatic system allows nanoparticles to stay for prolonged times inside the tumor.

Despite the advantages discussed above, the use of nanoparticles for drug delivery faces several limitations. First, tumor blood vessel leakiness could reduce tumor perfusion and therefore hinder nanoparticle delivery. As discussed in Chapter 4,

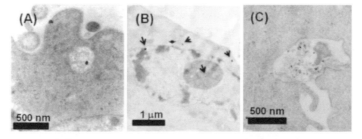

Figure 10.1 Nanoparticles can penetrate into the cell. Electron micrographs of fibroblasts treated with Ag nanoparticles. Arrows show the nanoparticles inside the cell. Image by Asharani et al., 2009, CC-BY-2.0 licence.

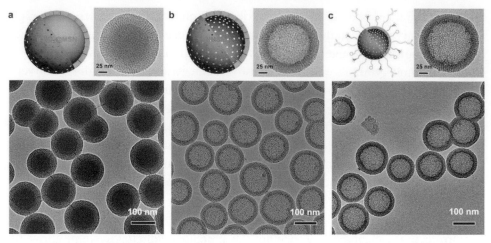

Figure 10.2 Nanoparticles can be used to encapsulate drugs. Here, for example, we show transmission electron micrographs of HMSN and HMSN-ZW800-TRC105 nanoparticles. Image by Chen et al., 2014a, CC-BY-4.0 licence.

tumors are associated with increased IFP, which could eliminate the pressure difference between the interior and exterior of the vessel, decreasing interstitial flow. Hence, nanoparticle transport would mostly occur by diffusion (Gerlowski and Jain, 1986; Nugent and Jain, 1984; Boucher and Jain, 1992), hindering the delivery of slowly diffusing large particles (Cabral et al., 2011; Stylianopoulos, 2013). It is worth noting that computational models of vascular tumor growth (Welter and Rieger, 2013), reviewed in Section 4.4, are currently used to corroborate these hypotheses and test strategies to optimize drug delivery. A second important emerging problem is that the efficacy of nanoparticles can be different in tumors that are rich in collagen and hyaluronic acid (e.g., fibrosarcomas, soft tissue sarcomas, pancreatic cancers) or in others having a moderate desmoplastic reaction (e.g., liver cancers) (Jain and Stylianopoulos, 2010).

In addition to improved drug transport, nanoparticles also offer direct strategies to eradicate tumors by inducing physical effects (e.g., magnetic, photothermal or photodynamic) on cancer cells after the activation of an external trigger by a laser or a magnetic field. Hyperthermia has been used to kill cancer cells but it is difficult to control the local temperature uniformly and precisely using traditional methods, leading to non-specific cell damage. Nanomaterials can be used to convert effectively an external energy input into heat. For example, magnetic nanoparticles can be used to dissipate energy into heat when stimulated by a rapidly oscillating magnetic field (Sasikala et al., 2015). Similarly, photothermal effects can be triggered in gold and carbon nanomaterials by deep-penetrating near-infrared light (von Maltzahn et al., 2009; Tao et al., 2014).

Radiotherapy uses high-energy radiation to kill cancer cells and is a common part of cancer therapy in the clinic. Radiation therapy is quite successful at eradicating cancer cells, but has adverse side effects ranging from damaged heart and lung tissue to infertility and, in some cases, development of a secondary cancer. Furthermore, cells in solid tumors are much more resistant to treatment due to the hypoxic nature of the intratumor environment. One approach to circumvent this significant challenge in radiotherapy is the development of nanocarriers to combine chemotherapeutics with radiosensitizers to synergistically increase the effectiveness of radiation therapy. Certain nanomaterials can be utilized as adjuvant radiosensitizers in combination with other therapies such as tumor ablation.

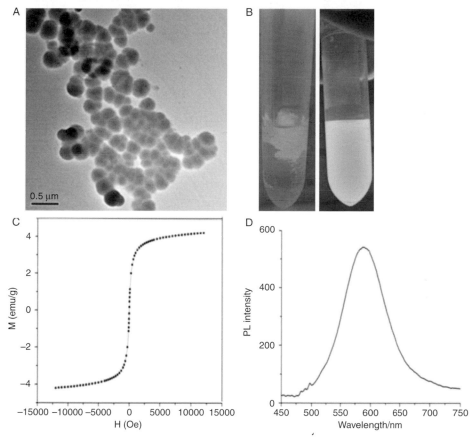

Figure 10.3 Magnetic nanoparticles can be used to kill cancer cells by applying a rapidly oscillating magnetic field to provoke hyperthermia. The image shows the characterization of fluorescent magnetic nanoparticles. A) Transmission electron microscope image. B) Image in water before and after amino-modification. C) Hysteresis curve. D) Photoluminescence spectra. Image by Li et al., 2015, CC-BY-3.0 licence.

A new therapeutic concept has been advocated in recent years: precision medicine. The idea of this approach is to tailor treatment and prevention by taking into account individual variability in genes, environment and lifestyle in order to maximize effectiveness. To achieve this goal it is fundamental to obtain more precise measurements of potential contributors, including molecular signatures from DNA sequencing technologies and environmental exposures or other information available from increasingly ubiquitous mobile devices. A precise understanding of molecular, environmental and behavioral profiles contributing to health and disease should lead to more accurate diagnoses and prevention strategies, better treatment selection, and the development of novel therapies. The main problem of current strategies is the rising costs of drug development and healthcare all over the world, suggesting that a new model of clinical care is needed. In addition, drug discovery has slowed down considerably, and only a small fraction of proposed medications are successfully translated into approved and prescribed therapies. In this context, contributions and ideas stemming from the physics of complex systems, including network science and big data analysis, could be highly innovative and significant.

10.2 Nutrition and Cancer

The literature that shows a relationship between nutrition and cancer is quite complex and often contradictory. Many studies provide, however, evidence that dietary components may affect the process of carcinogenesis. Isothiocyanates from cruciferous vegetables (cauliflower, cabbage and broccoli), diallyl sulfide (an organosulphur compound from garlic), isoflavone, phytosterole, folate, selenium, vitamin E, flavonoids and dietary fiber may reduce the risk of cancer, but how do they affect the behavior of the cell? Emerging evidence suggests that the protective effects from these dietary components can be mediated through epigenetic mechanisms. Of particular interest in this respect is the modulation of miRNA by nutrition. Essential nutrients and phytochemicals were shown to be able to modulate the expression of miRNA in different cancer cells and in other model systems.

Only a few studies examine the effects of various dietary patterns (e.g., Western diet) or alterations in macronutrient content (e.g., caloric restriction) on miRNA expression and function. A recent example is a study focusing on the effects of a Western diet, defined as a diet with increased animal fat and lower levels of cholecalciferol and calcium to approximate amounts consumed in Western diets (Newmark et al., 2001), on EGFR regulated miRNA in colonic tumorigenesis (Zhu et al., 2011). Earlier reports suggested that tumor promotion by a similar diet required EGFR signals, including MYC and K-Ras (Dougherty et al., 2009). The same investigators more recently found that EGFR suppressed miR-143 and miR-145 and that a Western diet would hinder the tumor suppressor effects of these miRNA by promoting colonic tumorigenesis and upregulating targets of these miRNA (i.e., MYC and K-Ras) (Zhu et al., 2011).

Additional in vivo studies examining the influence of macronutrients and/or dietary pattern on miRNA expression and function would likely provide promising insights into nutrition and cancer prevention. An example is the association of miRNA expression with various exposures, including alcohol intake, and clinical features associated with head and neck squamous cell carcinomas (Avissar et al., 2009): analyzing tumor tissues from 169 cases, the authors show that the expression of miR-375 increases significantly with alcohol consumption and is higher in tumors of pharyngeal and laryngeal origin when compared with oral tumors. Further studies will be important to elucidate possible association of dietary variables with circulating miRNA in individuals at high risk for certain cancers. Many essential nutrients are demonstrated to be involved in models of cancer prevention and risk (Ross and Davis, 2011) but the picture is still fragmented. In the next years it will be important to investigate how the different pieces are connected together. In this context, computational analysis of expression data and metabolic pathway simulations may open new avenues to our understanding.

11

Outlook on the Physics of Cancer: A New Interdisciplinary Area

In this book, we have tried to define the expanding boundaries of the relatively new field of the physics of cancer. Traditionally, physicists have contributed to cancer research mostly through the development of novel diagnostic and imaging tools. Things started to change in the last few years when physicists became more and more involved in trying to understand the roots of cancer and its development, bringing to the field their experience with quantitative modeling and data analysis. The basic idea is that cellular processes should ultimately obey the laws of physics: cell migration or mitosis occurs thanks to physical forces; tumors grow into tissues and are thus subject to mechanical and hydrodynamic forces. To understand these issues one needs to perform quantitative measurements and develop theoretical models as physicists have been doing for centuries.

In the last few years, cancer research witnessed the emergence of several promising new avenues deserving further investigation. Biology is currently undergoing a real revolution brought by the sheer growth of readily available quantitative data on all kind of biological processes in general and on cancer in particular. A considerable international effort is currently underway to assemble large databases of genetic mutations, transcriptomes and miRNA for all kinds of tumors from hundreds or sometimes thousands of patients. The ultimate goal of these efforts is to pave the way to a new type of personalized or precision medicine in which treatment will be tailored to the specific genetic and epigenetic features of each patient. Traditional training in biology is, however, often insufficient to deal with the mathematical and computational complexity associated with big data, which are instead the bread-and-butter of physicists. So while these big projects are not driven by physics, many of the people involved were trained in physics.

While genetic, transcriptomic, proteomic and metabolomic data are steadily accumulating in public databases, their interpretation is still a pressing challenge. The first problem stems from the fact that often data in different experiments are recorded using various methods with differences in normalization, formatting and notation. Hence it is often hard to treat and compare different data sets at the same time. Efforts are, however, underway to produce and assemble publicly available homogeneous databases for cancer, like TCGA (`https://tcga-data.nci.nih.gov/tcga/`). Studying large data sets could help answer fundamental questions about the emergence of tumors. Cancer is a multifactorial pathology that reflects the complex regulatory network inside the cell. Finding main hubs of this network, and the different ways it can function when perturbed, is a challenging task for the future.

Among the relevant and promising themes where we expect a contribution coming from physical sciences, we would also like to mention the role of chromatin conformations in gene regulation (Risca and Greenleaf, 2015). Cancer cells are characterized by wide chromatin alterations (see section 8.3), suggesting an important, but yet unclear, regulatory role of chromatin conformations (Reddy and Feinberg, 2013). Novel chromosome conformation capture techniques allow a precise map of chromatin topology to be obtained, and in particular of the location of contact points between chromatin domains (Giorgetti et al., 2014). A quantitative interpretation of three-dimensional DNA conformations would require the development of accurate large-scale numerical simulations (Dans et al., 2016). The problem is complex because of the multiple hierarchical scales involved (Gibcus and Dekker, 2013), from the small-scale behavior of single nucleotides that require methods based on quantum mechanics, such as *ab initio* molecular dynamics and quantum Montecarlo, to larger scales that can be approached by classical molecular dynamics or molecular mechanics that can overcome the typical timescale limitations of the former method. Atomistic methods, however, are unable to model large-scale chromatin topological features which require coarse-grained approaches such as polymer models (Dans et al., 2016).

The key role of the immune system in controlling each function of the organism is so important that understanding its regulation could help fight cancer. Indeed, a promising avenue for therapeutic intervention relies on strengthening the immune response against tumors or on weakening the mechanisms by which tumors evade the immune system. These interactions between immunity and cancer are very intriguing and complex topics which could benefit from quantitative methods and models.

This book aims to contribute to the training of a new generation of biologists and cancer researchers who should be able to combine the standard laboratory skills of biochemistry, cell biology and imaging with mathematical and computational tools for analysis and modeling. While we are still far from this goal, we are convinced that future cancer biology should not be completely removed from mathematical theories but should instead embrace them. Experiments will still remain informative but, as the amount of data grows, quantitative modeling and computational analysis will become imperative to disentangle its complexity. There is an enormous amount of knowledge to be gained from transforming a biological "cartoon" into a mathematical model. As was recently noted by Rob Phillips, in biological papers, theory, if present at all, is currently relegated to the last figure (Phillips, 2015). This is strikingly different in physics, where theory can not only appear in the first figure, but can span entire papers and even research fields. The suspicion against theorists is extraordinarily well rooted in biology: more than a century ago, the Nobel prizewinner Santiago Ramon y Cajal, in his advice to young investigators, warned against theorists, ranked alongside megalomaniacs and contemplators (Ramon y Cajal, 1897). It is slightly ironic that only a few years later, Albert Einstein, a theorist, was revered as the iconic physicist. One century later, it is probably the right time to overcome these distinctions and let theorists and experimentalists work together in biology.

While a physics training can be useful to tackle the complex problems posed by cancer research, it is by no means sufficient. There is an enormous body of sophisticated knowledge on the biological processes ruling cell behavior that cannot be ignored. The present book tries to distill the minimal and essential information needed by physicists to orient themselves in cancer research. We hope that we succeeded in this admittedly complicated endeavor.

References

Abdel Wahab, Siddig Ibrahim, Abdul, Ahmad Bustamam, Alzubairi, Adel Sharaf, Mohamed Elhassan, Manal and Mohan, Syam. 2009. In vitro ultramorphological assessment of apoptosis induced by zerumbone on (HeLa). *J Biomed Biotechnol*, **2009**, 769568.

Al-Hajj, Muhammad, Wicha, Max S, Benito-Hernandez, Adalberto, Morrison, Sean J and Clarke, Michael F. 2003. Prospective identification of tumorigenic breast cancer cells. *Proc Natl Acad Sci U S A*, **100**(7), 3983–8.

Alessandri, Kévin, Sarangi, Bibhu Ranjan, Gurchenkov, Vasily Valériévitch, Sinha, Bidisha, Kießling, Tobias Reinhold, Fetler, Luc, Rico, Felix, Scheuring, Simon, Lamaze, Christophe, Simon, Anthony et al. 2013. Cellular capsules as a tool for multi-cellular spheroid production and for investigating the mechanics of tumor progression in vitro. *Proceedings of the National Academy of Sciences*, **110**(37), 14843–8.

Alexiou, Panagiotis, Maragkakis, Manolis, Papadopoulos, Giorgos L, Reczko, Martin and Hatzigeorgiou, Artemis G. 2009. Lost in translation: an assessment and perspective for computational microRNA target identification. *Bioinformatics*, **25**(23), 3049–55.

Allen, Theresa M and Cullis, Pieter R. 2004. Drug delivery systems: entering the mainstream. *Science*, **303**(5665), 1818–22.

Amano, Takanori, Sagai, Tomoko, Tanabe, Hideyuki, Mizushina, Yoichi, Nakazawa, Hiromi and Shiroishi, Toshihiko. 2009. Chromosomal dynamics at the Shh locus: limb bud-specific differential regulation of competence and active transcription. *Dev Cell*, **16**(1), 47–57.

Anderson, Kristina, Lutz, Christoph, van Delft, Frederik W, Bateman, Caroline M, Guo, Yanping, Colman, Susan M, Kempski, Helena, Moorman, Anthony V, Titley, Ian, Swansbury, John, Kearney, Lyndal, Enver, Tariq and Greaves, Mel. 2011. Genetic variegation of clonal architecture and propagating cells in leukaemia. *Nature*, **469**(7330), 356–61.

Angelini, Thomas E, Hannezo, Edouard, Trepat, Xavier, Fredberg, Jeffrey J and Weitz, David A. 2010. Cell migration driven by cooperative substrate deformation patterns. *Phys Rev Lett*, **104**(16), 168104.

Angelini, Thomas E, Hannezo, Edouard, Trepat, Xavier, Marquez, Manuel, Fredberg, Jeffrey J and Weitz, David A. 2011. Glass-like dynamics of collective cell migration. *Proc Natl Acad Sci U S A*, **108**(12), 4714–9.

Antal, Tibor and Krapivsky, PL. 2010. Exact solution of a two-type branching process: clone size distribution in cell division kinetics. *Journal of Statistical Mechanics: Theory and Experiment*, **2010**(07), P07028.

Antonio, C, Ferby, I, Wilhelm, H., Jones, M. and Karsenti, E. 2000. Xkid, a chromokinesin required for chromosome alignment on the metaphase plate. *Cell*, **102**, 425–35.

Araujo, RP and McElwain, DLS. 2004. A history of the study of solid tumour growth: the contribution of mathematical modelling. *Bulletin of mathematical biology*, **66**(5), 1039–91.

Armitage, P and Doll, R. 1954. The age distribution of cancer and a multi-stage theory of carcinogenesis. *Br J Cancer*, **8**(1), 1–12.

Artandi, Steven E and DePinho, Ronald A. 2010. Telomeres and telomerase in cancer. *Carcinogenesis*, **31**(1), 9–18.

Asharani, PV, Hande, M Prakash and Valiyaveettil, Suresh. 2009. Anti-proliferative activity of silver nanoparticles. *BMC Cell Biol*, **10**, 65.

Ashkenazi, Rina, Gentry, Sara N and Jackson, Trachette L. 2008. Pathways to tumorigenesis–modeling mutation acquisition in stem cells and their progeny. *Neoplasia*, **10**(11), 1170–82.

Avissar, Michele, McClean, Michael D, Kelsey, Karl T and Marsit, Carmen J. 2009. MicroRNA expression in head and neck cancer associates with alcohol consumption and survival. *Carcinogenesis*, **30**(12), 2059–63.

Baeriswyl, Vanessa and Christofori, Gerhard. 2009. The angiogenic switch in carcinogenesis. *Semin Cancer Biol*, **19**(5), 329–37.

Bakhoum, Samuel F and Compton, Duane A. 2012. Chromosomal instability and cancer: a complex relationship with therapeutic potential. *J Clin Invest*, **122**(4), 1138–43.

Bakhoum, Samuel F, Genovese, Giulio and Compton, Duane A. 2009a. Deviant kinetochore microtubule dynamics underlie chromosomal instability. *Curr Biol*, **19**(22), 1937–42.

Bakhoum, Samuel F, Thompson, Sarah L, Manning, Amity L and Compton, Duane A. 2009b. Genome stability is ensured by temporal control of kinetochore-microtubule dynamics. *Nat Cell Biol*, **11**(1), 27–35.

Bancroft, James, Auckland, Philip, Samora, Catarina P and McAinsh, Andrew D. 2015. Chromosome congression is promoted by CENP-Q- and CENP-E-dependent pathways. *J Cell Sci*, **128**(1), 171–84.

Banerjee, Shiladitya, Utuje, Kazage JC and Marchetti, M Cristina. 2015. Propagating stress waves during epithelial expansion. *Phys Rev Lett*, **114**(22), 228101.

Banerji, S, Ni, J, Wang, SX, Clasper, S, Su, J, Tammi, R, Jones, M and Jackson, DG. 1999. LYVE-1, a new homologue of the CD44 glycoprotein, is a lymph-specific receptor for hyaluronan. *J Cell Biol*, **144**(4), 789–801.

Bao, Gang and Suresh, S. 2003. Cell and molecular mechanics of biological materials. *Nature Materials*, **2**(11), 715–25.

Baraldi, Massimiliano Maria, Alemi, Alexander A, Sethna, James P, Caracciolo, Sergio, La Porta, Caterina AM and Zapperi, Stefano. 2013. Growth and form of melanoma cell colonies. *Journal of Statistical Mechanics: Theory and Experiment*, **2013**(02), P02032.

Barisic, Marin, Aguiar, Paulo, Geley, Stephan and Maiato, Helder. 2014. Kinetochore motors drive congression of peripheral polar chromosomes by overcoming random arm-ejection forces. *Nat Cell Biol*, **16**(12), 1249–56.

Baron, JA and Sandler, RS. 2000. Nonsteroidal anti-inflammatory drugs and cancer prevention. *Annu Rev Med*, **51**, 511–23.

Bartha, K and Rieger, H. 2006. Vascular network remodeling via vessel cooption, regression and growth in tumors. *J Theor Biol*, **241**(4), 903–18.

Baxter, Laurence T and Jain, Rakesh K. 1989. Transport of fluid and macromolecules in tumors. I. Role of interstitial pressure and convection. *Microvascular Research*, **37**(1), 77–104.

Bazellières, Elsa, Conte, Vito, Elosegui-Artola, Alberto, Serra-Picamal, Xavier, Bintanel-Morcillo, María, Roca-Cusachs, Pere, Muñoz, José J, Sales-Pardo, Marta, Guimerà, Roger and Trepat, Xavier. 2015. Control of cell–cell forces and collective cell dynamics by the intercellular adhesome. *Nat Cell Biol*, **17**(4), 409–20.

Beerenwinkel, Niko, Antal, Tibor, Dingli, David, Traulsen, Arne, Kinzler, Kenneth W, Velculescu, Victor E, Vogelstein, Bert and Nowak, Martin A. 2007. Genetic progression and the waiting time to cancer. *PLoS Comput Biol*, **3**(11), e225.

Ben Amar, M, Chatelain, C and Ciarletta, P. 2011. Contour instabilities in early tumor growth models. *Phys Rev Lett*, **106**(14), 148101.

Berenblum, I and Shubik, P. 1949. An experimental study of the initiating stage of carcinogenesis, and a re-examination of the somatic cell mutation theory of cancer. *British Journal of Cancer*, **3**(1), 109–18.

Bergers, Gabriele and Benjamin, Laura E. 2003. Tumorigenesis and the angiogenic switch. *Nat Rev Cancer*, **3**(6), 401–10.

Bertalan, Zsolt, La Porta, Caterina AM, Maiato, Helder and Zapperi, Stefano. 2014. Conformational Mechanism for the Stability of Microtubule-Kinetochore Attachments. *Biophys. J.*, **107**, 289–300.

Bertalan, Zsolt, Budrikis, Zoe, La Porta, Caterina AM and Zapperi, Stefano. 2015. Role of the number of microtubules in chromosome segregation during cell division. *PLoS One*, **10**(10), e0141305.

Bhalla, US and Iyengar, R. 1999. Emergent properties of networks of biological signaling pathways. *Science*, **283**(5400), 381–7.

Bissell, MJ. 1999. Tumor plasticity allows vasculogenic mimicry, a novel form of angiogenic switch. A rose by any other name? *Am J Pathol*, **155**(3), 675–9.

Bittner, M, Meltzer, P, Chen, Y, Jiang, Y, Seftor, E, Hendrix, M, Radmacher, M, Simon, R, Yakhini, Z, Ben-Dor, A, Sampas, N, Dougherty, E, Wang, E, Marincola, F, Gooden, C, Lueders, J, Glatfelter, A, Pollock, P, Carpten, J, Gillanders, E, Leja, D, Dietrich, K, Beaudry, C, Berens, M, Alberts, D and Sondak, V. 2000. Molecular classification of cutaneous malignant melanoma by gene expression profiling. *Nature*, **406**(6795), 536–40.

Boiko, Alexander D, Razorenova, Olga V, van de Rijn, Matt, Swetter, Susan M, Johnson, Denise L, Ly, Daphne P, Butler, Paris D, Yang, George P, Joshua, Benzion, Kaplan, Michael J, Longaker, Michael T and Weissman, Irving L. 2010. Human melanoma-initiating cells express neural crest nerve growth factor receptor CD271. *Nature*, **466**(7302), 133–7.

Bond, Jacquelyn, Roberts, Emma, Springell, Kelly, Lizarraga, Sofia B, Lizarraga, Sophia, Scott, Sheila, Higgins, Julie, Hampshire, Daniel J, Morrison, Ewan E, Leal, Gabriella F, Silva, Elias O, Costa, Suzana MR, Baralle, Diana, Raponi, Michela, Karbani, Gulshan, Rashid, Yasmin, Jafri, Hussain, Bennett, Christopher, Corry, Peter, Walsh, Christopher A and Woods, C Geoffrey. 2005. A centrosomal mechanism involving CDK5RAP2 and CENPJ controls brain size. *Nat Genet*, **37**(4), 353–5.

Bonnet, D and Dick, JE. 1997. Human acute myeloid leukemia is organized as a hierarchy that originates from a primitive hematopoietic cell. *Nat Med*, **3**(7), 730–7.

Bornholdt, Stefan. 2008. Boolean network models of cellular regulation: prospects and limitations. *J R Soc Interface*, **5 Suppl 1**(Aug), S85–94.

Borsig, Lubor, Wong, Richard, Hynes, Richard O, Varki, Nissi M and Varki, Ajit. 2002. Synergistic effects of L- and P-selectin in facilitating tumor metastasis can involve non-mucin ligands and implicate leukocytes as enhancers of metastasis. *Proc Natl Acad Sci U S A*, **99**(4), 2193–8.

Bossel Ben-Moshe, Noa, Avraham, Roi, Kedmi, Merav, Zeisel, Amit, Yitzhaky, Assif, Yarden, Yosef and Domany, Eytan. 2012. Context-specific microRNA analysis: identification of functional microRNAs and their mRNA targets. *Nucleic Acids Res*, **40**(21), 10614–27.

Boucher, Yves and Jain, Rakesh K. 1992. Microvascular pressure is the principal driving force for interstitial hypertension in solid tumors: implications for vascular collapse. *Cancer Research*, **52**(18), 5110–4.

Boucher, Yves, Baxter, Laurence T and Jain, Rakesh K. 1990. Interstitial pressure gradients in tissue-isolated and subcutaneous tumors: implications for therapy. *Cancer Research*, **50**(15), 4478–84.

Boveri, T. 1903. *Übermehrpolige Mitosenals Mittelzur Analysedes Zellkerns*. Stuber.

Bozic, Ivana, Antal, Tibor, Ohtsuki, Hisashi, Carter, Hannah, Kim, Dewey, Chen, Sining, Karchin, Rachel, Kinzler, Kenneth W, Vogelstein, Bert and Nowak, Martin A. 2010. Accumulation of driver and passenger mutations during tumor progression. *Proc Natl Acad Sci U S A*, **107**, 18545–50.

Bravo, Rafael and Axelrod, David E. 2013. A calibrated agent-based computer model of stochastic cell dynamics in normal human colon crypts useful for in silico experiments. *Theor Biol Med Model*, **10**, 66.

Bronsert, P, Enderle-Ammour, K, Bader, M, Timme, S, Kuehs, M, Csanadi, A, Kayser, G, Kohler, I, Bausch, D, Hoeppner, J, Hopt, UT, Keck, T, Stickeler, E, Passlick, B, Schilling, O, Reiss, CP, Vashist, Y, Brabletz, T, Berger, J, Lotz, J, Olesch, J, Werner, M and Wellner, UF. 2014. Cancer cell invasion and EMT marker expression: a three-dimensional study of the human cancer-host interface. *J Pathol*, **234**(3), 410–22.

Brown, KS, Hill, CC, Calero, GA, Myers, CR, Lee, KH, Sethna, JP and Cerione, RA. 2004. The statistical mechanics of complex signaling networks: nerve growth factor signaling. *Phys Biol*, **1**(3-4), 184–95.

Brugues, Agusti, Anon, Ester, Conte, Vito, Veldhuis, Jim H. Gupta, Mukund, Colombelli, Julien, Munoz, Jose J, Brodland, G Wayne, Ladoux, Benoit and Trepat, Xavier. 2014. Forces driving epithelial wound healing. *Nat Phys*, **10**(9), 683–90.

Burrell, Rebecca A, McGranahan, Nicholas, Bartek, Jiri and Swanton, Charles. 2013a. The causes and consequences of genetic heterogeneity in cancer evolution. *Nature*, **501**(7467), 338–45.

Burrell, Rebecca A, McClelland, Sarah E, Endesfelder, David, Groth, Petra, Weller, Marie-Christine, Shaikh, Nadeem, Domingo, Enric, Kanu, Nnennaya, Dewhurst, Sally M, Gronroos, Eva, Chew, Su Kit, Rowan, Andrew J, Schenk, Arne, Sheffer, Michal, Howell, Michael, Kschischo, Maik, Behrens, Axel, Helleday, Thomas, Bartek, Jiri, Tomlinson, Ian P and Swanton, Charles. 2013b. Replication stress links structural and numerical cancer chromosomal instability. *Nature*, **494**(7438), 492–6.

Butler, James P, Tolić-Nørrelykke, Iva Marija, Fabry, Ben and Fredberg, Jeffrey J. 2002. Traction fields, moments, and strain energy that cells exert on their surroundings. *Am J Physiol Cell Physiol*, **282**(3), C595–605.

Cabral, H, Matsumoto, Y, Mizuno, K, Chen, Q, Murakami, M, Kimura, M, Terada, Y, Kano, M R, Miyazono, K, Uesaka, M, Nishiyama, N and Kataoka, K. 2011. Accumulation of sub-100 nm polymeric micelles in poorly permeable tumours depends on size. *Nat Nanotechnol*, **6**(12), 815–23.

Cahill, DP, Lengauer, C, Yu, J, Riggins, GJ, Willson, JK, Markowitz, SD, Kinzler, KW and Vogelstein, B. 1998. Mutations of mitotic checkpoint genes in human cancers. *Nature*, **392**(6673), 300–3.

Cai, S, O'Connell, CB, Khodjakov, A and Walczak, CE. 2009. Chromosome congression in the absence of kinetochore fibres. *Nat. Cell Biol.*, **11**, 832–838.

Califano, Joseph P, and Reinhart-King, Cynthia A. 2010. Exogenous and endogenous force regulation of endothelial cell behavior. *J Biomech*, **43**(1), 79–86.

Campisi, Judith. 2013. Aging, cellular senescence, and cancer. *Annu Rev Physiol*, **75**, 685–705.

Cane, S, Ye, AA, Luks-Morgan, SJ and Maresca, TJ. 2013. Elevated polar ejection forces stabilize kinetochore-microtubule attachments. *J. Cell Biol.*, **200**, 203.

Carmeliet, P, and Jain, RK. 2000. Angiogenesis in cancer and other diseases. *Nature*, **407**(6801), 249–57.

Carmeliet, Peter. 2005. VEGF as a key mediator of angiogenesis in cancer. *Oncology*, **69 Suppl 3**, 4–10.

Carmeliet, Peter and Jain, Rakesh K. 2011. Molecular mechanisms and clinical applications of angiogenesis. *Nature*, **473**(7347), 298–307.

Casciari, Joseph J, Sotirchos, Stratis V and Sutherland, Robert M. 1992. Variations in tumor cell growth rates and metabolism with oxygen concentration, glucose concentration, and extracellular pH. *Journal of Cellular Physiology*, **151**(2), 386–94.

Chacón, Jeremy M and Gardner, Melissa K 2013. Analysis and Modeling of Chromosome Congression During Mitosis in the Chemotherapy Drug Cisplatin. *Cell Mol Bioeng*, **6**(4), 406–17.

Chaffer, Christine L, Marjanovic, Nemanja D, Lee, Tony, Bell, George, Kleer, Celina G, Reinhardt, Ferenc, D'Alessio, Ana C, Young, Richard A and Weinberg, Robert A. 2013. Poised chromatin at the ZEB1 promoter enables breast cancer cell plasticity and enhances tumorigenicity. *Cell*, **154**(1), 61–74.

Chalut, Kevin J, Höpfler, Markus, Lautenschläger, Franziska, Boyde, Lars, Chan, Chii Jou, Ekpenyong, Andrew, Martinez-Arias, Alfonso and Guck, Jochen. 2012. Chromatin decondensation and nuclear softening accompany Nanog downregulation in embryonic stem cells. *Biophys J*, **103**(10), 2060–70.

Charafe-Jauffret, Emmanuelle, Ginestier, Christophe, Iovino, Flora, Wicinski, Julien, Cervera, Nathalie, Finetti, Pascal, Hur, Min Hee, Diebel, Mark F, Monville, Florence, Dutcher, Julie, Brown, Marty, Viens, Patrice, Xerri, Luc, Bertucci, François, Stassi, Giorgio, Dontu, Gabriela, Birnbaum, Daniel and Wicha, Max S. 2009. Breast cancer cell lines contain functional cancer stem cells with metastatic capacity and a distinct molecular signature. *Cancer Res*, **69**(4), 1302–13.

Charras, Guillaume T, Yarrow, Justin C, Horton, Mike A, Mahadevan, L and Mitchison, T J. 2005. Non-equilibration of hydrostatic pressure in blebbing cells. *Nature*, **435**(7040), 365–9.

Charras, Guillaume T, Hu, Chi-Kuo, Coughlin, Margaret and Mitchison, Timothy J. 2006. Reassembly of contractile actin cortex in cell blebs. *J Cell Biol*, **175**(3), 477–90.

Chatelain, C, Balois, T, Ciarletta, P and Amar, M Ben. 2011. Emergence of microstructural patterns in skin cancer: a phase separation analysis in a binary mixture. *New Journal of Physics*, **13**(11), 115013.

Chen, Duyu, Jiao, Yang and Torquato, Salvatore. 2014a. A cellular automaton model for tumor dormancy: emergence of a proliferative switch. *PLoS One*, **9**(10), e109934.

Chen, Feng, Hong, Hao, Shi, Sixiang, Goel, Shreya, Valdovinos, Hector F, Hernandez, Reinier, Theuer, Charles P, Barnhart, Todd E and Cai, Weibo. 2014b. Engineering of hollow mesoporous silica nanoparticles for remarkably enhanced tumor active targeting efficacy. *Sci Rep*, **4**, 5080.

Chen, Jian, Li, Yanjiao, Yu, Tzong-Shiue, McKay, Renée M, Burns, Dennis K, Kernie, Steven G and Parada, Luis F 2012. A restricted cell population propagates glioblastoma growth after chemotherapy. *Nature*, **488**(7412), 522–6.

Chen, Maureen Hong-Sing, Yip, George Wai-Cheong, Tse, Gary Man-Kit, Moriya, Takuya, Lui, Philip Chi-Wai, Zin, Mar-Lwin, Bay, Boon-Huat and Tan, Puay-Hoon. 2008. Expression of basal keratins and vimentin in breast cancers of young women correlates with adverse pathologic parameters. *Modern pathology*, **21**(10), 1183–1191.

Cheng, Gang, Tse, Janet, Jain, Rakesh K and Munn, Lance L. 2009. Micro-environmental mechanical stress controls tumor spheroid size and morphology by suppressing proliferation and inducing apoptosis in cancer cells. *PLoS one*, **4**(2), e4632.

Cherkezyan, Lusik, Stypula-Cyrus, Yolanda, Subramanian, Hariharan, White, Craig, Dela Cruz, Mart, Wali, Ramesh K, Goldberg, Michael J, Bianchi, Laura K, Roy, Hemant K and Backman, Vadim. 2014. Nanoscale changes in chromatin organization represent the initial steps of tumorigenesis: a transmission electron microscopy study. *BMC Cancer*, **14**, 189.

Cheung, AMS, Wan, TSK, Leung, CKJ, Chan, LYY, Huang, H, Kwong, YL, Liang, R and Leung, AYH. 2007. Aldehyde dehydrogenase activity in leukemic blasts defines a subgroup of acute myeloid leukemia with adverse prognosis and superior NOD/SCID engrafting potential. *Leukemia*, **21**(7), 1423–30.

Cheung, Kevin J, Gabrielson, Edward, Werb, Zena and Ewald, Andrew J. 2013. Collective invasion in breast cancer requires a conserved basal epithelial program. *Cell*, **155**(7), 1639–51.

Cheung, Kevin J, Padmanaban, Veena, Silvestri, Vanesa, Schipper, Koen, Cohen, Joshua D, Fairchild, Amanda N, Gorin, Michael A, Verdone, James E, Pienta, Kenneth J, Bader, Joel S and Ewald, Andrew J. 2016. Polyclonal breast cancer metastases arise from collective dissemination of keratin 14-expressing tumor cell clusters. *Proc Natl Acad Sci U S A*, **113**(7), E854–63.

Cho, Robert W, Wang, Xinhao, Diehn, Maximilian, Shedden, Kerby, Chen, Grace Y, Sherlock, Gavin, Gurney, Austin, Lewicki, John and Clarke, Michael F. 2008. Isolation and molecular characterization of cancer stem cells in MMTV-Wnt-1 murine breast tumors. *Stem Cells*, **26**(2), 364–71.

Chrzanowska-Wodnicka, Magdalena and Burridge, Keith. 1996. Rho-stimulated contractility drives the formation of stress fibers and focal adhesions. *The Journal of Cell Biology*, **133**(6), 1403–1415.

Chu, Yeh-Shiu, Dufour, Sylvie, Thiery, Jean Paul, Perez, Eric and Pincet, Frederic. 2005. Johnson-Kendall-Roberts theory applied to living cells. *Phys Rev Lett*, **94**(2), 028102.

Chute, John P, Muramoto, Garrett G, Whitesides, John, Colvin, Michael, Safi, Rachid, Chao, Nelson J and McDonnell, Donald P. 2006. Inhibition of aldehyde dehydrogenase and retinoid signaling induces the expansion of human hematopoietic stem cells. *Proc Natl Acad Sci U S A*, **103**(31), 11707–12.

Ciliberto, Andrea, Novak, Béla, and Tyson, John J. 2005. Steady states and oscillations in the p53/Mdm2 network. *Cell Cycle*, **4**(3), 488–93.

Cimini, Daniela, Wan, Xiaohu, Hirel, Christophe B, and Salmon, E.D. 2006. Aurora kinase promotes turnover of kinetochore microtubules to reduce chromosome segregation errors. *Curr. Biol.*, **16**, 1711–18.

Civelekoglu-Scholey, G, Sharp, D, Mogilner, A and Scholey, JM. 2006. Model of chromosome motility in drosophila embryos: adaptation of a general mechanism for rapid mitosis. *Biophys. J.*, **90**, 3966.

Clayton, Elizabeth, Doupé, David P, Klein, Allon M, Winton, Douglas J, Simons, Benjamin D and Jones, Philip H. 2007. A single type of progenitor cell maintains normal epidermis. *Nature*, **446**(7132), 185–9.

Codling, Edward A, Plank, Michael J and Benhamou, Simon. 2008. Random walk models in biology. *J R Soc Interface*, **5**(25), 813–34.

Coffey, JC, Wang, JH, Smith, MJF, Bouchier-Hayes, D, Cotter, TG and Redmond, H P. 2003. Excisional surgery for cancer cure: therapy at a cost. *Lancet Oncol*, **4**(12), 760–8.

Collado, Manuel and Serrano, Manuel. 2010. Senescence in tumours: evidence from mice and humans. *Nat Rev Cancer*, **10**(1), 51–7.

Cooper, G.M. 2000. *The Cell: A Molecular Approach. 2nd edition*. Sunderland (MA): Sinauer Associates.

Corti, Stefania, Locatelli, Federica, Papadimitriou, Dimitra, Donadoni, Chiara, Salani, Sabrina, Del Bo, Roberto, Strazzer, Sandra, Bresolin, Nereo and Comi, Giacomo P. 2006. Identification of a primitive brain-derived neural stem cell population based on aldehyde dehydrogenase activity. *Stem Cells*, **24**(4), 975–85.

Cross, Sarah E, Jin, Yu-Sheng, Rao, Jianyu and Gimzewski, James K. 2007. Nanomechanical analysis of cells from cancer patients. *Nature Nanotechnology*, **2**(12), 780–3.

Dai, Xin and Tan, Chalet. 2015. Combination of microRNA therapeutics with small-molecule anticancer drugs: mechanism of action and co-delivery nanocarriers. *Adv Drug Deliv Rev*, **81**(Jan), 184–97.

Dans, Pablo D, Walther, Jürgen, Gómez, Hansel and Orozco, Modesto. 2016. Multiscale simulation of DNA. *Curr Opin Struct Biol*, **37**(Apr), 29–45.

Davoli, Teresa, Denchi, Eros Lazzerini and de Lange, Titia. 2010. Persistent telomere damage induces bypass of mitosis and tetraploidy. *Cell*, **141**(1), 81–93.

De Bock, Katrien, Georgiadou, Maria, Schoors, Sandra, Kuchnio, Anna, Wong, Brian W, Cantelmo, Anna Rita, Quaegebeur, Annelies, Ghesquière, Bart, Cauwenberghs, Sandra, Eelen, Guy, Phng, Li-Kun, Betz, Inge, Tembuyser, Bieke, Brepoels, Katleen, Welti, Jonathan, Geudens, Ilse, Segura, Inmaculada, Cruys, Bert, Bifari, Franscesco, Decimo, Ilaria, Blanco, Raquel, Wyns, Sabine, Vangindertael, Jeroen, Rocha, Susana, Collins, Russel T, Munck, Sebastian, Daelemans, Dirk, Imamura, Hiromi, Devlieger, Roland, Rider, Mark, Van Veldhoven, Paul P, Schuit, Frans, Bartrons, Ramon, Hofkens, Johan, Fraisl, Peter, Telang, Sucheta, Deberardinis, Ralph J, Schoonjans, Luc, Vinckier, Stefan, Chesney, Jason, Gerhardt, Holger, Dewerchin, Mieke and Carmeliet, Peter. 2013. Role of PFKFB3-driven glycolysis in vessel sprouting. *Cell*, **154**(3), 651–63.

de Jong, Hidde. 2002. Modeling and simulation of genetic regulatory systems: a literature review. *J Comput Biol*, **9**(1), 67–103.

de Pillis, Lisette G, Radunskaya, Ami E and Wiseman, Charles L. 2005. A validated mathematical model of cell-mediated immune response to tumor growth. *Cancer Res*, **65**(17), 7950–8.

Deisboeck, Thomas S, Wang, Zhihui, Macklin, Paul and Cristini, Vittorio. 2011. Multiscale cancer modeling. *Annu Rev Biomed Eng*, **13**(Aug), 127–55.

Dejana, Elisabetta, Orsenigo, Fabrizio, Molendini, Cinzia, Baluk, Peter and McDonald, Donald M. 2009. Organization and signaling of endothelial cell-to-cell junctions in various regions of the blood and lymphatic vascular trees. *Cell Tissue Res*, **335**(1), 17–25.

Dembo, Micah and Wang, Yu-Li. 1999. Stresses at the cell-to-substrate interface during locomotion of fibroblasts. *Biophysical Journal*, **76**(4), 2307–16.

Demou, Zoe N. 2010. Gene expression profiles in 3D tumor analogs indicate compressive strain differentially enhances metastatic potential. *Annals of Biomedical Engineering*, **38**(11), 3509–20.

Demou, Zoe N and Hendrix, Mary JC. 2008. Microgenomics profile the endogenous angiogenic phenotype in subpopulations of aggressive melanoma. *J Cell Biochem*, **105**(2), 562–73.

Denais, Celine and Lammerding, Jan. 2014. Nuclear mechanics in cancer. *Adv Exp Med Biol*, **773**, 435–70.

Denais, Celine M, Gilbert, Rachel M, Isermann, Philipp, McGregor, Alexandra L, te Lindert, Mariska, Weigelin, Bettina, Davidson, Patricia M, Friedl, Peter, Wolf, Katarina and Lammerding, Jan. 2016. Nuclear envelope rupture and repair during cancer cell migration. *Science*, **352**(6283), 353–8.

Dieterich, Peter, Klages, Rainer, Preuss, Roland and Schwab, Albrecht. 2008. Anomalous dynamics of cell migration. *Proc Natl Acad Sci U S A*, **105**(2), 459–63.

Ding, Yi-tao, Kumar, Shant and Yu, De-cai. 2008. The role of endothelial progenitor cells in tumour vasculogenesis. *Pathobiology*, **75**(5), 265–73.

DiResta, Gene R, Lee, Jongbin, Healey, John H, Levchenko, Andrey, Larson, Steven M and Arbit, Ehud. 2000. "Artificial lymphatic system": a new approach to reduce interstitial hypertension and increase blood flow, pH and pO2 in solid tumors. *Annals of Biomedical Engineering*, **28**(5), 543–555.

Domnitz, Sarah B, Wagenbach, Michael, Decarreau, Justin and Wordeman, Linda. 2012. MCAK activity at microtubule tips regulates spindle microtubule length to promote robust kinetochore attachment. *J Cell Biol*, **197**(2), 231–7.

Dou, Jun, Pan, Meng, Wen, Ping, Li, Yating, Tang, Quan, Chu, Lili, Zhao, Fengshu, Jiang, Chuilian, Hu, Weihua, Hu, Kai and Gu, Ning. 2007. Isolation and identification of cancer stem-like cells from murine melanoma cell lines. *Cell Mol Immunol*, **4**(6), 467–72.

Dougherty, Urszula, Cerasi, Dario, Taylor, Ieva, Kocherginsky, Masha, Tekin, Ummuhan, Badal, Shamiram, Aluri, Lata, Sehdev, Amikar, Cerda, Sonia, Mustafi, Reba, Delgado, Jorge, Joseph, Loren, Zhu, Hongyan, Hart, John, Threadgill, David, Fichera, Alessandro and Bissonnette, Marc. 2009. Epidermal growth factor receptor is required for colonic tumor promotion by dietary fat in the azoxymethane/dextran sulfate sodium model: roles of transforming growth factor-alpha and PTGS2. *Clin Cancer Res*, **15**(22), 6780–9.

Downward, Julian. 2003. Targeting RAS signalling pathways in cancer therapy. *Nat Rev Cancer*, **3**(1), 11–22.

Doyle, Andrew D, Wang, Francis W, Matsumoto, Kazue and Yamada, Kenneth M. 2009. One-dimensional topography underlies three-dimensional fibrillar cell migration. *J Cell Biol*, **184**(4), 481–90.

Dranoff, Glenn. 2004. Cytokines in cancer pathogenesis and cancer therapy. *Nat Rev Cancer*, **4**(1), 11–22.

Drasdo, D, Kree, R and McCaskill, JS. 1995. Monte Carlo approach to tissue-cell populations. *Physical Review E*, **52**(6), 6635–57.

Drasdo, Dirk and Hoehme, Stefan. 2005. A single-cell-based model of tumor growth in vitro: monolayers and spheroids. *Physical Biology*, **2**(3), 133.

Drasdo, Dirk and Hoehme, Stefan. 2012. Modeling the impact of granular embedding media, and pulling versus pushing cells on growing cell clones. *New Journal of Physics*, **14**(5), 055025.

Drasdo, Dirk, Hoehme, Stefan and Block, Michael. 2007. On the role of physics in the growth and pattern formation of multi-cellular systems: What can we learn from individual-cell based models? *Journal of Statistical Physics*, **128**(1-2), 287–345.

Drier, Yotam, Sheffer, Michal and Domany, Eytan. 2013. Pathway-based personalized analysis of cancer. *Proc Natl Acad Sci U S A*, **110**(16), 6388–93.

Driessens, Gregory, Beck, Benjamin, Caauwe, Amélie, Simons, Benjamin D and Blanpain, Cédric. 2012. Defining the mode of tumour growth by clonal analysis. *Nature*, **488**(7412), 527–30.

Ebata, N, Nodasaka, Y, Sawa, Y, Yamaoka, Y, Makino, S, Totsuka, Y and Yoshida, S. 2001. Desmoplakin as a specific marker of lymphatic vessels. *Microvasc Res*, **61**(1), 40–8.

Engler, Adam J, Griffin, Maureen A, Sen, Shamik, Bönnemann, Carsten G, Sweeney, H Lee and Discher, Dennis E. 2004. Myotubes differentiate optimally on substrates with tissue-like stiffness pathological implications for soft or stiff microenvironments. *The Journal of Cell Biology*, **166**(6), 877–87.

Engler, Adam J, Sen, Shamik, Sweeney, H Lee and Discher, Dennis E. 2006. Matrix elasticity directs stem cell lineage specification. *Cell*, **126**(4), 677–89.

Ertych, Norman, Stolz, Ailine, Stenzinger, Albrecht, Weichert, Wilko, Kaulfuß, Silke, Burfeind, Peter, Aigner, Achim, Wordeman, Linda and Bastians, Holger. 2014. Increased microtubule assembly rates influence chromosomal instability in colorectal cancer cells. *Nat Cell Biol*, **16**(8), 779–91.

Fabry, B, Maksym, GN, Butler, JP, Glogauer, M, Navajas, D and Fredberg, JJ. 2001. Scaling the microrheology of living cells. *Phys Rev Lett*, **87**(14), 148102.

Fackler, Oliver T and Grosse, Robert. 2008. Cell motility through plasma membrane blebbing. *J Cell Biol*, **181**(6), 879–84.

Fang, Dong, Nguyen, Thiennga K, Leishear, Kim, Finko, Rena, Kulp, Angela N, Hotz, Susan, Van Belle, Patricia A, Xu, Xiaowei, Elder, David E and Herlyn, Meenhard. 2005. A tumorigenic subpopulation with stem cell properties in melanomas. *Cancer Res*, **65**(20), 9328–37.

Farokhzad, Omid C and Langer, Robert. 2009. Impact of nanotechnology on drug delivery. *ACS Nano*, **3**(1), 16–20.

Fearon, ER and Vogelstein, B. 1990. A genetic model for colorectal tumorigenesis. *Cell*, **61**(5), 759–67.

Ferrara, Napoleone. 2009. Vascular endothelial growth factor. *Arterioscler Thromb Vasc Biol*, **29**(6), 789–91.

Folkman, Judah. 2002. Role of angiogenesis in tumor growth and metastasis. *Semin Oncol*, **29**(6 Suppl 16), 15–8.

Freyer, James P and Sutherland, Robert M. 1986. Regulation of growth saturation and development of necrosis in EMT6/Ro multicellular spheroids by the glucose and oxygen supply. *Cancer Research*, **46**(7), 3504–12.

Friedl, Peter and Gilmour, Darren. 2009. Collective cell migration in morphogenesis, regeneration and cancer. *Nat Rev Mol Cell Biol*, **10**(7), 445–57.

Friedl, Peter, Wolf, Katarina and Lammerding, Jan. 2011. Nuclear mechanics during cell migration. *Curr Opin Cell Biol*, **23**(1), 55–64.

Friedl, Peter, Locker, Joseph, Sahai, Erik and Segall, Jeffrey E. 2012. Classifying collective cancer cell invasion. *Nat Cell Biol*, **14**(8), 777–83.

Fritsch, Anatol, Hockel, Michael, Kiessling, Tobias, Nnetu, Kenechukwu David, Wetzel, Franziska, Zink, Mareike and Kas, Josef A. 2010. Are biomechanical changes necessary for tumour progression? *Nat Phys*, **6**(10), 730–2.

Fukumura, Dai and Jain, Rakesh K. 2007. Tumor microenvironment abnormalities: causes, consequences, and strategies to normalize. *Journal of Cellular Biochemistry*, **101**(4), 937–49.

Fuller, Brian G, Lampson, Michael A, Foley, Emily A, Rosasco-Nitcher, Sara, Le, Kim V, Tobelmann, Page, Brautigan, David L, Stukenberg, P Todd and Kapoor, Tarun M. 2008. Midzone activation of aurora B in anaphase produces an intracellular phosphorylation gradient. *Nature*, **453**(7198), 1132–6.

Fumia, Herman F and Martins, Marcelo L. 2013. Boolean network model for cancer pathways: predicting carcinogenesis and targeted therapy outcomes. *PLoS One*, **8**(7), e69008.

Fussner, Eden, Ching, Reagan W and Bazett-Jones, David P. 2011. Living without 30nm chromatin fibers. *Trends Biochem Sci*, **36**(1), 1–6.

Galle, Jorg, Loeffler, Markus and Drasdo, Dirk. 2005. Modeling the effect of deregulated proliferation and apoptosis on the growth dynamics of epithelial cell populations in vitro. *Biophysical Journal*, **88**(1), 62–75.

Ganguly, Anutosh, Yang, Hailing and Cabral, Fernando. 2011. Overexpression of mitotic centromere-associated Kinesin stimulates microtubule detachment and confers resistance to paclitaxel. *Mol Cancer Ther*, **10**(6), 929–37.

Gardner, Melissa K, Pearson, Chad G, Sprague, Brian L, Zarzar, Ted R, Bloom, Kerry, Salmon, ED and Odde, David J. 2005. Tension-dependent regulation of microtubule dynamics at kinetochores can explain metaphase congression in yeast. *Mol. Biol. Cell*, **16**, 3764–75.

Garra, Brian Stephen. 2007. Imaging and estimation of tissue elasticity by ultrasound. *Ultrasound Quarterly*, **23**(4), 255–68.

Georges, Penelope C, Miller, William J, Meaney, David F, Sawyer, Evelyn S and Janmey, Paul A. 2006. Matrices with compliance comparable to that of brain tissue select neuronal over glial growth in mixed cortical cultures. *Biophysical Journal*, **90**(8), 3012–18.

Gerlowski, LE and Jain, RK. 1986. Microvascular permeability of normal and neoplastic tissues. *Microvasc Res*, **31**(3), 288–305.

Geva-Zatorsky, Naama, Rosenfeld, Nitzan, Itzkovitz, Shalev, Milo, Ron, Sigal, Alex, Dekel, Erez, Yarnitzky, Talia, Liron, Yuvalal, Polak, Paz, Lahav, Galit and Alon, Uri. 2006. Oscillations and variability in the p53 system. *Mol Syst Biol*, **2**, 2006.0033.

Giancotti, FG and Ruoslahti, E. 1999. Integrin signaling. *Science*, **285**(5430), 1028–32.

Gibcus, Johan H and Dekker, Job. 2013. The hierarchy of the 3D genome. *Mol Cell*, **49**(5), 773–82.

Gieni, Randall S and Hendzel, Michael J. 2008. Mechanotransduction from the ECM to the genome: are the pieces now in place? *Journal of Cellular Biochemistry*, **104**(6), 1964–1987.

Gillespie, Daniel T. 1976. A general method for numerically simulating the stochastic time evolution of coupled chemical reactions. *Journal of Computational Physics*, **22**(4), 403–34.

Ginestier, Christophe, Hur, Min Hee, Charafe-Jauffret, Emmanuelle, Monville, Florence, Dutcher, Julie, Brown, Marty, Jacquemier, Jocelyne, Viens, Patrice, Kleer, Celina G, Liu, Suling, Schott, Anne, Hayes, Dan, Birnbaum, Daniel, Wicha, Max S and Dontu, Gabriela. 2007. ALDH1 is a marker of normal and malignant human mammary stem cells and a predictor of poor clinical outcome. *Cell Stem Cell*, **1**(5), 555–67.

Ginsberg, Mark H, Du, Xiaoping and Plow, Edward F. 1992. Inside-out integrin signalling. *Current Opinion in Cell Biology*, **4**(5), 766–771.

Giorgetti, Luca, Galupa, Rafael, Nora, Elphège P, Piolot, Tristan, Lam, France, Dekker, Job, Tiana, Guido and Heard, Edith. 2014. Predictive polymer modeling reveals coupled fluctuations in chromosome conformation and transcription. *Cell*, **157**(4), 950–63.

Glaser, Kevin J, Felmlee, Joel P, Manduca, Armando, Kannan Mariappan, Yogesh and Ehman, Richard L. 2006. Stiffness-weighted magnetic resonance imaging. *Magnetic Resonance in Medicine*, **55**(1), 59–67.

Glunčić, Matko, Maghelli, Nicola, Krull, Alexander, Krstić, Vladimir, Ramunno-Johnson, Damien, Pavin, Nenad and Tolić, Iva M. 2015. Kinesin-8 motors improve nuclear centering by promoting microtubule catastrophe. *Phys Rev Lett*, **114**(7), 078103.

Gnjatic, Sacha, Cao, Yanran, Reichelt, Uta, Yekebas, Emre F, Nölker, Christina, Marx, Andreas H, Erbersdobler, Andreas, Nishikawa, Hiroyoshi, Hildebrandt, York, Bartels, Katrin, Horn, Christiane, Stahl, Tanja, Gout, Ivan, Filonenko, Valeriy, Ling, Khoon-Lin, Cerundolo, Vincenzo, Luetkens, Tim, Ritter, Gerd, Friedrichs, Kay, Leuwer, Rudolf, Hegewisch-Becker, Susanna, Izbicki, Jakob R, Bokemeyer, Carsten, Old, Lloyd J and Atanackovic, Djordje. 2010. NY-CO-58/KIF2C is overexpressed in a variety of solid tumors and induces frequent T cell responses in patients with colorectal cancer. *Int J Cancer*, **127**(2), 381–93.

Gordon, VD, Valentine, MT, Gardel, ML, Andor-Ardo, D, Dennison, S, Bogdanov, AA, Weitz, DA and Deisboeck, TS. 2003. Measuring the mechanical stress induced by an expanding multicellular tumor system: a case study. *Experimental Cell Research*, **289**(1), 58–66.

Gould, SJ and Eldredge, N. 1993. Punctuated equilibrium comes of age. *Nature*, **366**(6452), 223–7.

Gov, NS. 2014. *Cell and Matrix Mechanics*. CRC Press. Pages 219–38.

Greaves, Mel and Maley, Carlo C. 2012. Clonal evolution in cancer. *Nature*, **481**(7381), 306–13.

Greco, Francesca and Vicent, María J. 2009. Combination therapy: opportunities and challenges for polymer-drug conjugates as anticancer nanomedicines. *Adv Drug Deliv Rev*, **61**(13), 1203–13.

Griffon-Etienne, Geneviève, Boucher, Yves, Brekken, Christian, Suit, Herman D and Jain, Rakesh K. 1999. Taxane-induced apoptosis decompresses blood vessels and lowers interstitial fluid pressure in solid tumors clinical implications. *Cancer Research*, **59**(15), 3776–3782.

Guck, Jochen, Schinkinger, Stefan, Lincoln, Bryan, Wottawah, Falk, Ebert, Susanne, Romeyke, Maren, Lenz, Dominik, Erickson, Harold M, Ananthakrishnan, Revathi, Mitchell, Daniel, et al. 2005. Optical deformability as an inherent cell marker for testing malignant transformation and metastatic competence. *Biophysical Journal*, **88**(5), 3689–98.

Gumbiner, BM. 2000. Regulation of cadherin adhesive activity. *J Cell Biol*, **148**(3), 399–404.

Gundem, Gunes, Van Loo, Peter, Kremeyer, Barbara, Alexandrov, Ludmil B, Tubio, Jose M C, Papaemmanuil, Elli, Brewer, Daniel S, Kallio, Heini M L, Högnäs, Gunilla, Annala, Matti, Kivinummi, Kati, Goody, Victoria, Latimer, Calli, O'Meara, Sarah, Dawson, Kevin J, Isaacs, William, Emmert-Buck, Michael R, Nykter, Matti, Foster, Christopher, Kote-Jarai, Zsofia, Easton, Douglas, Whitaker, Hayley C, ICGC Prostate UK Group, Neal, David E, Cooper, Colin S, Eeles, Rosalind A, Visakorpi, Tapio, Campbell, Peter J, McDermott, Ultan, Wedge, David C and Bova, G Steven. 2015. The evolutionary history of lethal metastatic prostate cancer. *Nature*, **520**(7547), 353–7.

Gupta, Piyush B, Fillmore, Christine M, Jiang, Guozhi, Shapira, Sagi D, Tao, Kai, Kuperwasser, Charlotte and Lander, Eric S. 2011. Stochastic state transitions give rise to phenotypic equilibrium in populations of cancer cells. *Cell*, **146**(4), 633–44.

Gutenkunst, Ryan N, Waterfall, Joshua J, Casey, Fergal P, Brown, Kevin S, Myers, Christopher R and Sethna, James P. 2007. Universally sloppy parameter sensitivities in systems biology models. *PLoS Comput Biol*, **3**(10), 1871–8.

Ha, Sung Chul, Lowenhaupt, Ky, Rich, Alexander, Kim, Yang-Gyun and Kim, Kyeong Kyu. 2005. Crystal structure of a junction between B-DNA and Z-DNA reveals two extruded bases. *Nature*, **437**(7062), 1183–6.

Hadnagy, Annamaria, Gaboury, Louis, Beaulieu, Raymond and Balicki, Danuta. 2006. SP analysis may be used to identify cancer stem cell populations. *Exp Cell Res*, **312**(19), 3701–10.

Haga, Hisashi, Irahara, Chikako, Kobayashi, Ryo, Nakagaki, Toshiyuki and Kawabata, Kazushige. 2005. Collective movement of epithelial cells on a collagen gel substrate. *Biophys J*, **88**(3), 2250–6.

Hanahan, D and Folkman, J. 1996. Patterns and emerging mechanisms of the angiogenic switch during tumorigenesis. *Cell*, **86**(3), 353–64.

Hanahan, Douglas and Weinberg, Robert A. 2000. The hallmarks of cancer. *Cell*, **100**(1), 57–70.

Hanahan, Douglas and Weinberg, Robert A. 2011. Hallmarks of cancer: the next generation. *Cell*, **144**(5), 646–74.

Hardy, Katharine M, Kirschmann, Dawn A, Seftor, Elisabeth A, Margaryan, Naira V, Postovit, Lynne-Marie, Strizzi, Luigi and Hendrix, Mary JC. 2010. Regulation of the embryonic morphogen Nodal by Notch4 facilitates manifestation of the aggressive melanoma phenotype. *Cancer Res*, **70**(24), 10340–50.

Harris, AK, Wild, P and Stopak, D. 1980. Silicone rubber substrata: a new wrinkle in the study of cell locomotion. *Science*, **208**(4440), 177–9.

Harris, TE. 1989. *The Theory of Branching Processes*. Dover, New York.

Harrison, Hannah, Farnie, Gillian, Brennan, Keith R and Clarke, Robert B. 2010. Breast cancer stem cells: something out of notching? *Cancer Res*, **70**(22), 8973–6.

Hayflick, L and Moorhead, PS. 1961. The serial cultivation of human diploid cell strains. *Experimental Cell Research*, **25**(3), 585–621.

Hayflick, Leonard. 2007. Entropy explains aging, genetic determinism explains longevity, and undefined terminology explains misunderstanding both. *PLoS Genet*, **3**(12), e220.

Heldin, Carl-Henrik, Rubin, Kristofer, Pietras, Kristian and Östman, Arne. 2004. High interstitial fluid pressure—an obstacle in cancer therapy. *Nature Reviews Cancer*, **4**(10), 806–13.

Helmlinger, Gabriel, Netti, Paolo A, Lichtenbeld, Hera C, Melder, Robert J and Jain, Rakesh K. 1997. Solid stress inhibits the growth of multicellular tumor spheroids. *Nature Biotechnology*, **15**(8), 778–83.

Hendrix, Mary JC, Seftor, Richard EB, Seftor, Elisabeth A, Gruman, Lynn M, Lee, Lisa ML, Nickoloff, Brian J, Miele, Lucio, Sheriff, Don D, and Schatteman, Gina C. 2002. Transendothelial function of human metastatic melanoma cells: role of the microenvironment in cell-fate determination. *Cancer Res*, **62**(3), 665–8.

Hendrix, Mary JC, Seftor, Elisabeth A, Hess, Angela R and Seftor, Richard EB. 2003. Molecular plasticity of human melanoma cells. *Oncogene*, **22**(20), 3070–5.

Herzmark, Paul, Campbell, Kyle, Wang, Fei, Wong, Kit, El-Samad, Hana, Groisman, Alex and Bourne, Henry R. 2007. Bound attractant at the leading vs. the trailing edge determines chemotactic prowess. *Proc Natl Acad Sci U S A*, **104**(33), 13349–54.

Hill, TL. 1985. Theoretical problems related to the attachment of microtubules to kinetochores. *Proc Natl Acad Sci U S A*, **82**(13), 4404–8.

Holash, J, Maisonpierre, PC, Compton, D, Boland, P, Alexander, CR, Zagzag, D, Yancopoulos, GD and Wiegand, SJ. 1999. Vessel cooption, regression, and growth in tumors mediated by angiopoietins and VEGF. *Science*, **284**(5422), 1994–8.

Holcombe, Mike, Adra, Salem, Bicak, Mesude, Chin, Shawn, Coakley, Simon, Graham, Alison I, Green, Jeffrey, Greenough, Chris, Jackson, Duncan, Kiran, Mariam,

MacNeil, Sheila, Maleki-Dizaji, Afsaneh, McMinn, Phil, Pogson, Mark, Poole, Robert, Qwarnstrom, Eva, Ratnieks, Francis, Rolfe, Matthew D, Smallwood, Rod, Sun, Tao and Worth, David. 2012. Modelling complex biological systems using an agent-based approach. *Integr Biol (Camb)*, **4**(1), 53–64.

Holy, TE and Leibler, S. 1995. Dynamic instability of microtubules as an efficient way to search in space. *PNAS*, **91**, 5682–5.

Huang, Sui and Ingber, Donald E. 2005. Cell tension, matrix mechanics, and cancer development. *Cancer Cell*, **8**(3), 175–176.

Huelsken, J and Birchmeier, W. 2001. New aspects of Wnt signaling pathways in higher vertebrates. *Curr Opin Genet Dev*, **11**(5), 547–53.

Hung, Liang-Yi, Chen, Hua-Ling, Chang, Ching-Wen, Li, Bor-Ran and Tang, Tang K. 2004. Identification of a novel microtubule-destabilizing motif in CPAP that binds to tubulin heterodimers and inhibits microtubule assembly. *Mol Biol Cell*, **15**(6), 2697–706.

Hunter, Andrew W, Caplow, Michael, Coy, David L, Hancock, William O, Diez, Stefan, Wordeman, Linda and Howard, Jonathon. 2003. The kinesin-related protein MCAK is a microtubule depolymerase that forms an ATP-hydrolyzing complex at microtubule ends. *Mol Cell*, **11**(2), 445–57.

Hynes, Richard O. 1992. Integrins: versatility, modulation, and signaling in cell adhesion. *Cell*, **69**(1), 11–25.

Iglesias, Pablo A and Devreotes, Peter N. 2008. Navigating through models of chemotaxis. *Curr Opin Cell Biol*, **20**(1), 35–40.

Ilina, Olga and Friedl, Peter. 2009. Mechanisms of collective cell migration at a glance. *J Cell Sci*, **122**(Pt 18), 3203–8.

Ingber, Donald E, Madri, Joseph A and Jamieson, James D. 1981. Role of basal lamina in neoplastic disorganization of tissue architecture. *Proceedings of the National Academy of Sciences*, **78**(6), 3901–5.

Itzkovitz, Shalev, Blat, Irene C, Jacks, Tyler, Clevers, Hans and van Oudenaarden, Alexander. 2012. Optimality in the development of intestinal crypts. *Cell*, **148**(3), 608–19.

Jain, Rakesh K. 1988. Determinants of tumor blood flow: a review. *Cancer Research*, **48**(10), 2641–58.

Jain, Rakesh K and Stylianopoulos, Triantafyllos. 2010. Delivering nanomedicine to solid tumors. *Nat Rev Clin Oncol*, **7**(11), 653–4.

Jain, RK and Duda, DG. 2004. Vascular and interstitial biology of tumors. *Clinical Oncology*, Abeleff M, Armitage J, Niederhuber J, Kastan M, McKenna G (eds). Elsevier, Philadelphia, PA, 153–172.

Janet, M Tse, Cheng, Gang, Tyrrell, James A, Wilcox-Adelman, Sarah A, Boucher, Yves, Jain, Rakesh K and Munn, Lance L. 2012. Mechanical compression drives cancer cells toward invasive phenotype. *Proceedings of the National Academy of Sciences*, **109**(3), 911–16.

Janmey, Paul A, Euteneuer, Ursula, Traub, Peter and Schliwa, Manfred. 1991. Viscoelastic properties of vimentin compared with other filamentous biopolymer networks. *The Journal of Cell Biology*, **113**(1), 155–60.

Jiao, Yang and Torquato, Salvatore. 2011. Emergent behaviors from a cellular automaton model for invasive tumor growth in heterogeneous microenvironments. *PLoS Comput Biol*, **7**(12), e1002314.

Jiao, Yang and Torquato, Salvatore. 2012. Diversity of dynamics and morphologies of invasive solid tumors. *AIP Adv*, **2**(1), 11003.

Jiao, Yang and Torquato, Salvatore. 2013. Evolution and morphology of microenvironment-enhanced malignancy of three-dimensional invasive solid tumors. *Phys Rev E Stat Nonlin Soft Matter Phys*, **87**(5), 052707.

Joanny, Jean-François and Prost, Jacques. 2009. Active gels as a description of the actin-myosin cytoskeleton. *HFSP J*, **3**(2), 94–104.

Joglekar, AP and Hunt, AJ. 2002. A simple, mechanistic model for directional instability during mitotic chromosome movement. *Biophys. J.*, **83**, 42–58.

Joukov, V, Pajusola, K, Kaipainen, A, Chilov, D, Lahtinen, I, Kukk, E, Saksela, O, Kalkki-nen, N and Alitalo, K. 1996. A novel vascular endothelial growth factor, VEGF-C, is a ligand for the Flt4 (VEGFR-3) and KDR (VEGFR-2) receptor tyrosine kinases. *EMBO J*, **15**(7), 1751.

Kaipainen, A, Korhonen, J, Mustonen, T, van Hinsbergh, VW, Fang, GH, Dumont, D, Breitman, M and Alitalo, K. 1995. Expression of the fms-like tyrosine kinase 4 gene becomes restricted to lymphatic endothelium during development. *Proc Natl Acad Sci U S A*, **92**(8), 3566–70.

Kam, Yoonseok and Quaranta, Vito. 2009. Cadherin-bound beta-catenin feeds into the Wnt pathway upon adherens junctions dissociation: evidence for an intersection between beta-catenin pools. *PLoS One*, **4**(2), e4580.

Kantoff, Philip W, Higano, Celestia S, Shore, Neal D, Berger, E Roy, Small, Eric J, Penson, David F, Redfern, Charles H, Ferrari, Anna C, Dreicer, Robert, Sims, Robert B, Xu, Yi, Frohlich, Mark W, Schellhammer, Paul F, and IMPACT Study Investigators. 2010. Sipuleucel-T immunotherapy for castration-resistant prostate cancer. *N Engl J Med*, **363**(5), 411–22.

Kapoor, TM, Lampson, MA, Hergert, P., Cameron, L., Cimini, D., Salmon, ED, McEwen, BF and Khodjakov, A. 2006. Chromosomes can congress to the metaphase plate before biorientation. *Science*, **311**, 388–91.

Kauffman, Stuart, Peterson, Carsten, Samuelsson, Björn and Troein, Carl. 2003. Random Boolean network models and the yeast transcriptional network. *Proc Natl Acad Sci U S A*, **100**(25), 14796–9.

Kenific, Candia M, Thorburn, Andrew and Debnath, Jayanta. 2010. Autophagy and metastasis: another double-edged sword. *Curr Opin Cell Biol*, **22**(2), 241–5.

Keshet, Gilmor I, Goldstein, Itamar, Itzhaki, Orit, Cesarkas, Karen, Shenhav, Liraz, Yakire-vitch, Arkadi, Treves, Avraham J, Schachter, Jacob, Amariglio, Ninette and Rechavi, Gideon. 2008. MDR1 expression identifies human melanoma stem cells. *Biochem Biophys Res Commun*, **368**(4), 930–6.

Kessenbrock, Kai, Plaks, Vicki and Werb, Zena. 2010. Matrix metalloproteinases: regulators of the tumor microenvironment. *Cell*, **141**(1), 52–67.

Kessler, David A, Austin, Robert H, and Levine, Herbert. 2014. Resistance to chemotherapy: patient variability and cellular heterogeneity. *Cancer Res*, **74**(17), 4663–70.

Khalil, Antoine A and Friedl, Peter. 2010. Determinants of leader cells in collective cell migration. *Integr Biol (Camb)*, **2**(11-12), 568–74.

Kim, YG, Kim, PS, Herbert, A and Rich, A. 1997. Construction of a Z-DNA-specific restriction endonuclease. *Proc Natl Acad Sci U S A*, **94**(24), 12875–9.

Kimmel, M. and Axelrod, DE. 2002. *Branching Processes in Biology*. Springer, New York.

Kimmel, Marek and Axelrod, David E. 1991. Unequal cell division, growth regulation and colony size of mammalian cells: A mathematical model and analysis of experimental data. *Journal of Theoretical Biology*, **153**(2), 157–80.

Kirschner, D and Panetta, JC. 1998. Modeling immunotherapy of the tumor-immune interaction. *J Math Biol*, **37**(3), 235–52.

Kirschner, M and Mitchison, T. J. 1986. Beyond self-assembly: from microtubules to morphogenesis. *Cell*, **45**, 329–42.

Klein, Walter M, Wu, Bryan P, Zhao, Shuping, Wu, Hong, Klein-Szanto, Andres JP and Tahan, Steven R. 2007. Increased expression of stem cell markers in malignant melanoma. *Mod Pathol*, **20**(1), 102–7.

Kloor, Matthias, Michel, Sara and von Knebel Doeberitz, Magnus. 2010. Immune evasion of microsatellite unstable colorectal cancers. *Int J Cancer*, **127**(5), 1001–10.

Knudson, AG. 2001. Two genetic hits (more or less) to cancer. *Nat Rev Cancer*, **1**(2), 157–62.

Knudson, Jr, AG. 1971. Mutation and cancer: statistical study of retinoblastoma. *Proc Natl Acad Sci U S A*, **68**(4), 820–3.

Koch, Thorsten M, Münster, Stefan, Bonakdar, Navid, Butler, James P and Fabry, Ben. 2012. 3D Traction forces in cancer cell invasion. *PLoS One*, **7**(3), e33476.

Kohlmaier, Gregor, Loncarek, Jadranka, Meng, Xing, McEwen, Bruce F, Mogensen, Mette M, Spektor, Alexander, Dynlacht, Brian D, Khodjakov, Alexey and Gönczy, Pierre. 2009. Overly long centrioles and defective cell division upon excess of the SAS-4-related protein CPAP. *Curr Biol*, **19**(12), 1012–8.

Koike, C, McKee, TD, Pluen, A, Ramanujan, S, Burton, K, Munn, LL, Boucher, Y and Jain, RK. 2002. Solid stress facilitates spheroid formation: potential involvement of hyaluronan. *British Journal of Cancer*, **86**(6), 947–53.

Kollmannsberger, Philip and Fabry, Ben. 2011. Linear and Nonlinear Rheology of Living Cells. *Annual Review of Materials Research*, **41**(1), 75–97.

König, Peter, Braunfeld, Michael B, Sedat, John W and Agard, David A. 2007. The three-dimensional structure of in vitro reconstituted Xenopus laevis chromosomes by EM tomography. *Chromosoma*, **116**(4), 349–72.

Korolev, Nikolay, Lyubartsev, Alexander P and Nordenskiöld, Lars. 2010. Cation-induced polyelectrolyte-polyelectrolyte attraction in solutions of DNA and nucleosome core particles. *Adv Colloid Interface Sci*, **158**(1-2), 32–47.

Kraning-Rush, Casey M, Califano, Joseph P and Reinhart-King, Cynthia A. 2012. Cellular traction stresses increase with increasing metastatic potential. *PLoS One*, **7**(2), e32572.

Krijger, Peter HL and de Laat, Wouter. 2013. Identical cells with different 3D genomes; cause and consequences? *Curr Opin Genet Dev*, **23**(2), 191–6.

Kukk, E, Lymboussaki, A, Taira, S, Kaipainen, A, Jeltsch, M, Joukov, V and Alitalo, K. 1996. VEGF-C receptor binding and pattern of expression with VEGFR-3 suggests a role in lymphatic vascular development. *Development*, **122**(12), 3829–37.

La Porta, CAM. 2011. Transdifferentiation of cancer stem cells into endothelial cells. *Central European Journal of Biology*, **6**, 850–1.

La Porta, CAM, Ghilardi, Anna, Pasini, Maria, Laurson, Lasse, Alava, Mikko J, Zapperi, Stefano and Ben Amar, Martine. 2015a. Osmotic stress affects functional properties of human melanoma cell lines. *Eur. Phys. J. Plus*, **130**(4), 64.

La Porta, Caterina. 2009. Cancer stem cells: lessons from melanoma. *Stem Cell Rev*, **5**(1), 61–5.

La Porta, Caterina AM and Zapperi, Stefano. 2013. Human breast and melanoma cancer stem cells biomarkers. *Cancer Lett*, **338**(1), 69–73.

La Porta, Caterina AM, Ghilardi, Anna, Pasini, Maria, Laurson, Lasse, Alava, Mikko J, Zapperi, Stefano and Amar, Martine Ben. 2015b. Osmotic stress affects functional properties of human melanoma cell lines. *The European Physical Journal Plus*, **130**(4), 1–15.

La Porta, Caterina AM, Zapperi, Stefano and Sethna, James P. 2012. Senescent cells in growing tumors: population dynamics and cancer stem cells. *PLoS Comput Biol*, **8**(1), e1002316.

Lahav, Galit, Rosenfeld, Nitzan, Sigal, Alex, Geva-Zatorsky, Naama, Levine, Arnold J, Elowitz, Michael B and Alon, Uri. 2004. Dynamics of the p53-Mdm2 feedback loop in individual cells. *Nat Genet*, **36**(2), 147–50.

Lampson, MA and Cheeseman, IM. 2011. Sensing centromere tension: Aurora B and the regulation of kinetochore function. *Trends Cell Biol.*, **21**, 133–38.

Landau, DL and Lifishitz, EM. 1970. *Theory of Elasticity*. Pergamon Press.

Lane, DP. 1992. Cancer. p53, guardian of the genome. *Nature*, **358**(6381), 15–6.

Lange, Janina R and Fabry, Ben. 2013. Cell and tissue mechanics in cell migration. *Exp Cell Res*, **319**(16), 2418–23.

Langer, R. 1998. Drug delivery and targeting. *Nature*, **392**(6679 Suppl), 5–10.

Lautscham, Lena A, Kämmerer, Christoph, Lange, Janina R, Kolb, Thorsten, Mark, Christoph, Schilling, Achim, Strissel, Pamela L, Strick, Reiner, Gluth, Caroline, Rowat, Amy C, Metzner, Claus and Fabry, Ben. 2015. Migration in confined 3D environments is determined by a combination of adhesiveness, nuclear volume, contractility, and cell stiffness. *Biophys J*, **109**(5), 900–13.

LaVan, David A, McGuire, Terry and Langer, Robert. 2003. Small-scale systems for in vivo drug delivery. *Nat Biotechnol*, **21**(10), 1184–91.

Lawo, Steffen, Hasegan, Monica, Gupta, Gagan D and Pelletier, Laurence. 2012. Subdiffraction imaging of centrosomes reveals higher-order organizational features of pericentriolar material. *Nat Cell Biol*, **14**(11), 1148–58.

Lee, Alvin J X, Endesfelder, David, Rowan, Andrew J, Walther, Axel, Birkbak, Nicolai J, Futreal, P Andrew, Downward, Julian, Szallasi, Zoltan, Tomlinson, Ian P M, Howell, Michael, Kschischo, Maik and Swanton, Charles. 2011. Chromosomal instability confers intrinsic multidrug resistance. *Cancer Res*, **71**(5), 1858–70.

Lee, D-S, Rieger, H and Bartha, K. 2006. Flow correlated percolation during vascular remodeling in growing tumors. *Phys Rev Lett*, **96**(5), 058104.

Lee, Ethan, Salic, Adrian, Krüger, Roland, Heinrich, Reinhart and Kirschner, Marc W. 2003. The roles of APC and Axin derived from experimental and theoretical analysis of the Wnt pathway. *PLoS Biol*, **1**(1), E10.

Lee, Mi Kyeong and Nikodem, Vera M. 2004. Differential role of ERK in cAMP-induced Nurr1 expression in N2A and C6 cells. *Neuroreport*, **15**(1), 99–102.

Legant, Wesley R, Miller, Jordan S, Blakely, Brandon L, Cohen, Daniel M, Genin, Guy M and Chen, Christopher S. 2010. Measurement of mechanical tractions exerted by cells in three-dimensional matrices. *Nat Methods*, **7**(12), 969–71.

Lekka, M, Laidler, P, Gil, D, Lekki, J, Stachura, Z and Hrynkiewicz, AZ. 1999. Elasticity of normal and cancerous human bladder cells studied by scanning force microscopy. *European Biophysics Journal*, **28**(4), 312–16.

Lengauer, C, Kinzler, KW, and Vogelstein, B. 1998. Genetic instabilities in human cancers. *Nature*, **396**(6712), 643–9.

Leong, Kevin G, Wang, Bu-Er, Johnson, Leisa and Gao, Wei-Qiang. 2008. Generation of a prostate from a single adult stem cell. *Nature*, **456**(7223), 804–8.

Lev Bar-Or, R, Maya, R, Segel, LA, Alon, U, Levine, AJ and Oren, M. 2000. Generation of oscillations by the p53-Mdm2 feedback loop: a theoretical and experimental study. *Proc Natl Acad Sci U S A*, **97**(21), 11250–5.

Levental, Kandice R, Yu, Hongmei, Kass, Laura, Lakins, Johnathon N, Egeblad, Mikala, Erler, Janine T, Fong, Sheri FT, Csiszar, Katalin, Giaccia, Amato, Weninger, Wolfgang, et al. 2009. Matrix crosslinking forces tumor progression by enhancing integrin signaling. *Cell*, **139**(5), 891–906.

Levine, Herbert, Kessler, David A and Rappel, Wouter-Jan. 2006. Directional sensing in eukaryotic chemotaxis: a balanced inactivation model. *Proc Natl Acad Sci U S A*, **103**(26), 9761–6.

Li, Chao, Ruan, Jing, Yang, Meng, Pan, Fei, Gao, Guo, Qu, Su, Shen, You-Lan, Dang, Yong-Jun, Wang, Kan, Jin, Wei-Lin and Cui, Da-Xiang. 2015. Human induced pluripotent stem cells labeled with fluorescent magnetic nanoparticles for targeted imaging and hyperthermia therapy for gastric cancer. *Cancer Biol Med*, **12**(3), 163–74.

Li, Hangwen and Tang, Dean G. 2011. Prostate cancer stem cells and their potential roles in metastasis. *J Surg Oncol*, **103**(6), 558–62.

Li, Weifeng, Nordenskiöld, Lars, Zhou, Ruhong and Mu, Yuguang. 2014. Conformation-dependent DNA attraction. *Nanoscale*, **6**(12), 7085–92.

Li, Y, Yu, W, Liang, Y and Zhu, X. 2007. Kinetochore dynein generates a poleward pulling force to facilitate congression and full chromosome alignment. *Cell Res.*, **17**, 701–12.

Liotta, LA, Saidel, MG and Kleinerman, J. 1976. The significance of hematogenous tumor cell clumps in the metastatic process. *Cancer Res*, **36**(3), 889–94.

Liu, Dan, Vader, Gerben, Vromans, Martijn JM, Lampson, Michael A and Lens, Susanne MA. 2009. Sensing chromosome bi-orientation by spatial separation of aurora B kinase from kinetochore substrates. *Science*, **323**(5919), 1350–3.

Liu, Kai, Yuan, Yuan, Huang, Jianyong, Wei, Qiong, Pang, Mingshu, Xiong, Chunyang and Fang, Jing. 2013. Improved-throughput traction microscopy based on fluorescence micropattern for manual microscopy. *PLoS One*, **8**(8), e70122.

Liverpool, Tanniemola B and Marchetti, M Cristina. 2003. Instabilities of isotropic solutions of active polar filaments. *Phys Rev Lett*, **90**(13), 138102.

Livshits, Anna, Git, Anna, Fuks, Garold, Caldas, Carlos and Domany, Eytan. 2015. Pathway-based personalized analysis of breast cancer expression data. *Mol Oncol*, **9**(7), 1471–83.

López-Otín, Carlos, Blasco, Maria A, Partridge, Linda, Serrano, Manuel and Kroemer, Guido. 2013. The hallmarks of aging. *Cell*, **153**(6), 1194–1217.

Lowe, Martin, and Barr, Francis A. 2007. Inheritance and biogenesis of organelles in the secretory pathway. *Nat Rev Mol Cell Biol*, **8**(6), 429–39.

Lowengrub, JS, Frieboes, HB, Jin, F, Chuang, Y-L, Li, X, Macklin, P, Wise, SM and Cristini, V. 2010. Nonlinear modelling of cancer: bridging the gap between cells and tumours. *Nonlinearity*, **23**(1), R1–R9.

Luna, Coralia, Li, Guorong, Huang, Jianyong, Qiu, Jianming, Wu, Jing, Yuan, Fan, Epstein, David L and Gonzalez, Pedro. 2012. Regulation of trabecular meshwork cell contraction and intraocular pressure by miR-200c. *PLoS One*, **7**(12), e51688.

Mac Gabhann, Feilim and Popel, Aleksander S. 2008. Systems biology of vascular endothelial growth factors. *Microcirculation*, **15**(8), 715–38.

Maeda, Hiroshi. 2012. Vascular permeability in cancer and infection as related to macromolecular drug delivery, with emphasis on the EPR effect for tumor-selective drug targeting. *Proc Jpn Acad Ser B Phys Biol Sci*, **88**(3), 53–71.

Magidson, Valentin, O'Connell, Christopher B, Lončarek, Jadranka, Paul, Raja, Mogilner, Alex and Khodjakov, Alexey. 2011. The spatial arrangement of chromosomes during prometaphase facilitates spindle assembly. *Cell*, **146**(4), 555–67.

Makrilia, Nektaria, Kollias, Anastasios, Manolopoulos, Leonidas and Syrigos, Kostas. 2009. Cell adhesion molecules: role and clinical significance in cancer. *Cancer Invest*, **27**(10), 1023–37.

Mallet, DG and De Pillis, LG. 2006. A cellular automata model of tumor-immune system interactions. *J Theor Biol*, **239**(3), 334–50.

Manibog, Kristine, Li, Hui, Rakshit, Sabyasachi and Sivasankar, Sanjeevi. 2014. Resolving the molecular mechanism of cadherin catch bond formation. *Nat Commun*, **5**, 3941.

Marchetti, MC, Joanny, JF, Ramaswamy, S, Liverpool, TB, Prost, J, Rao, Madan and Simha, R Aditi. 2013. Hydrodynamics of soft active matter. *Rev. Mod. Phys.*, **85**(Jul), 1143–89.

Marjanovic, Nemanja D, Weinberg, Robert A and Chaffer, Christine L. 2013. Cell plasticity and heterogeneity in cancer. *Clin Chem*, **59**(1), 168–79.

Martinez-Sanchez, Aida and Murphy, Chris L. 2013. MicroRNA target identification—experimental approaches. *Biology*, **2**(1), 189–205.

Matioli, G, Niewisch, H and Vogel, H. 1970. Stochastic stem cell renewal. *Rev Eur Etud Clin Biol*, **15**(1), 20–2.

Matos, Irina, Pereira, António J, Lince-Faria, Mariana, Cameron, Lisa A, Salmon, Edward D and Maiato, Helder. 2009. Synchronizing chromosome segregation by flux-dependent force equalization at kinetochores. *J. Cell Biol.*, **186**, 11–26.

McBeath, Rowena, Pirone, Dana M, Nelson, Celeste M, Bhadriraju, Kiran and Chen, Christopher S. 2004. Cell shape, cytoskeletal tension, and RhoA regulate stem cell lineage commitment. *Developmental Cell*, **6**(4), 483–95.

McDonald, Donald M and Choyke, Peter L. 2003. Imaging of angiogenesis: from microscope to clinic. *Nat Med*, **9**(6), 713–25.

McGregor, Alexandra Lynn, Hsia, Chieh-Ren and Lammerding, Jan. 2016. Squish and squeeze-the nucleus as a physical barrier during migration in confined environments. *Curr Opin Cell Biol*, **40**(Jun), 32–40.

McIntosh, JR, Cande, WZ and Snyder, JA. 1975. Structure and physiology of the mammalian mitotic spindle. *Soc Gen Physiol Ser*, **30**, 31–76.

Medema, Jan Paul and Vermeulen, Louis. 2011. Microenvironmental regulation of stem cells in intestinal homeostasis and cancer. *Nature*, **474**(7351), 318–26.

Merkel, Rudolf, Kirchgessner, Norbert, Cesa, Claudia M and Hoffmann, Bernd. 2007. Cell force microscopy on elastic layers of finite thickness. *Biophys J*, **93**(9), 3314–23.

Merlo, Lauren M F, Pepper, John W, Reid, Brian J and Maley, Carlo C. 2006. Cancer as an evolutionary and ecological process. *Nat Rev Cancer*, **6**(12), 924–35.

Metzner, Claus, Mark, Christoph, Steinwachs, Julian, Lautscham, Lena, Stadler, Franz and Fabry, Ben. 2015. Superstatistical analysis and modelling of heterogeneous random walks. *Nat Commun*, **6**, 7516.

Michor, Franziska. 2008. Mathematical models of cancer stem cells. *J Clin Oncol*, **26**(17), 2854–61.

Michor, Franziska, Hughes, Timothy P, Iwasa, Yoh, Branford, Susan, Shah, Neil P, Sawyers, Charles L and Nowak, Martin A. 2005. Dynamics of chronic myeloid leukaemia. *Nature*, **435**(7046), 1267–7.

Mitchison, TJ, Charras, TG and Mahadevan, L. 2008. Implications of a poroelastic cytoplasm for the dynamics of animal cell shape. *Semin Cell Dev Biol*, **19**(3), 215–23.

Moeendarbary, Emad, Valon, Léo, Fritzsche, Marco, Harris, Andrew R, Moulding, Dale A, Thrasher, Adrian J, Stride, Eleanor, Mahadevan, L and Charras, Guillaume T. 2013. The cytoplasm of living cells behaves as a poroelastic material. *Nat Mater*, **12**(3), 253–61.

Montecalvo, Angela, Larregina, Adriana T, Shufesky, William J, Stolz, Donna Beer, Sullivan, Mara LG, Karlsson, Jenny M, Baty, Catherine J, Gibson, Gregory A, Erdos, Geza, Wang, Zhiliang, Milosevic, Jadranka, Tkacheva, Olga A, Divito, Sherrie J, Jordan, Rick, Lyons-Weiler, James, Watkins, Simon C and Morelli, Adrian E. 2012. Mechanism of transfer of functional microRNAs between mouse dendritic cells via exosomes. *Blood*, **119**(3), 756–66.

Montel, Fabien, Delarue, Morgan, Elgeti, Jens, Malaquin, Laurent, Basan, Markus, Risler, Thomas, Cabane, Bernard, Vignjevic, Danijela, Prost, Jacques, Cappello, Giovanni, et al. 2011. Stress clamp experiments on multicellular tumor spheroids. *Physical Review Letters*, **107**(18), 188102.

Montel, Fabien, Delarue, Morgan, Elgeti, Jens, Vignjevic, Danijela, Cappello, Giovanni and Prost, Jacques. 2012. Isotropic stress reduces cell proliferation in tumor spheroids. *New Journal of Physics*, **14**(5), 055008.

Monzani, Elena, Facchetti, Floriana, Galmozzi, Enrico, Corsini, Elena, Benetti, Anna, Cavazzin, Chiara, Gritti, Angela, Piccinini, Andrea, Porro, Danilo, Santinami, Mario, Invernici, Gloria, Parati, Eugenio, Alessandri, Giulio and La Porta, Caterina A M. 2007. Melanoma contains CD133 and ABCG2 positive cells with enhanced tumourigenic potential. *Eur J Cancer*, **43**(5), 935–46.

Monzani, Elena, Bazzotti, Riccardo, Perego, Carla and La Porta, Caterina AM. 2009. AQP1 is not only a water channel: it contributes to cell migration through Lin7/beta-catenin. *PLoS One*, **4**(7), e6167.

Moore, GE, Sandberg, AA and Watne, AL. 1960. The comparative size and structure of tumor cells and clumps in the blood, bone marrow, and tumor imprints. *Cancer*, **13**, 111–7.

Moran, Patrick Alfred Pierce. 1958. Random processes in genetics. Pages 60–71 of: *Mathematical Proceedings of the Cambridge Philosophical Society*, vol. 54. Cambridge University Press.

Muinonen-Martin, Andrew J, Veltman, Douwe M, Kalna, Gabriela and Insall, Robert H. 2010. An improved chamber for direct visualisation of chemotaxis. *PLoS One*, **5**(12), e15309.

Muller, HJ. 1930. Radiation and Genetics. *The American Naturalist*, **64**(692), 220–251.

Muller, HJ. 1964. The relation of recombination to mutational advance. *Mutat Res*, **106**(May), 2–9.

Nagy, Janice A, Chang, Sung-Hee, Shih, Shou-Ching, Dvorak, Ann M and Dvorak, Harold F. 2010. Heterogeneity of the tumor vasculature. *Semin Thromb Hemost*, **36**(3), 321–31.

Nair, Smita, Boczkowski, David, Moeller, Benjamin, Dewhirst, Mark, Vieweg, Johannes and Gilboa, Eli. 2003. Synergy between tumor immunotherapy and antiangiogenic therapy. *Blood*, **102**(3), 964–71.

Nakamura, Y, Tanaka, F, Haraguchi, N, Mimori, K, Matsumoto, T, Inoue, H, Yanaga, K and Mori, M. 2007. Clinicopathological and biological significance of mitotic centromere-associated kinesin overexpression in human gastric cancer. *Br J Cancer*, **97**(4), 543–9.

Navin, Nicholas, Kendall, Jude, Troge, Jennifer, Andrews, Peter, Rodgers, Linda, McIndoo, Jeanne, Cook, Kerry, Stepansky, Asya, Levy, Dan, Esposito, Diane, Muthuswamy, Lakshmi, Krasnitz, Alex, McCombie, W Richard, Hicks, James and Wigler, Michael. 2011. Tumour evolution inferred by single-cell sequencing. *Nature*, **472**(7341), 90–4.

Netti, Paolo A, Baxter, Laurence T, Boucher, Yves, Skalak, Richard and Jain, Rakesh K. 1995. Time-dependent behavior of interstitial fluid pressure in solid tumors: implications for drug delivery. *Cancer Research*, **55**(22), 5451–8.

Netti, Paolo A, Berk, David A, Swartz, Melody A, Grodzinsky, Alan J and Jain, Rakesh K. 2000. Role of extracellular matrix assembly in interstitial transport in solid tumors. *Cancer Research*, **60**(9), 2497–503.

Newmark, HL, Yang, K, Lipkin, M, Kopelovich, L, Liu, Y, Fan, K and Shinozaki, H. 2001. A Western-style diet induces benign and malignant neoplasms in the colon of normal C57Bl/6 mice. *Carcinogenesis*, **22**(11), 1871–5.

Ng, Mei Rosa, Besser, Achim, Danuser, Gaudenz and Brugge, Joan S. 2012. Substrate stiffness regulates cadherin-dependent collective migration through myosin-II contractility. *J Cell Biol*, **199**(3), 545–63.

Nielsen, M-B, Christensen, Soren T and Hoffmann, Else K. 2008. Effects of osmotic stress on the activity of MAPKs and PDGFR-β-mediated signal transduction in NIH-3T3 fibroblasts. *American Journal of Physiology-Cell Physiology*, **294**(4), C1046–C1055.

Niethammer, Andreas G, Xiang, Rong, Becker, Jürgen C, Wodrich, Harald, Pertl, Ursula, Karsten, Gabriele, Eliceiri, Brian P and Reisfeld, Ralph A. 2002. A DNA vaccine against VEGF receptor 2 prevents effective angiogenesis and inhibits tumor growth. *Nat Med*, **8**(12), 1369–75.

Nishino, Yoshinori, Eltsov, Mikhail, Joti, Yasumasa, Ito, Kazuki, Takata, Hideaki, Takahashi, Yukio, Hihara, Saera, Frangakis, Achilleas S, Imamoto, Naoko, Ishikawa, Tetsuya and Maeshima, Kazuhiro. 2012. Human mitotic chromosomes consist predominantly of irregularly folded nucleosome fibres without a 30-nm chromatin structure. *EMBO J*, **31**(7), 1644–53.

Nora, Elphège P, Lajoie, Bryan R, Schulz, Edda G, Giorgetti, Luca, Okamoto, Ikuhiro, Servant, Nicolas, Piolot, Tristan, van Berkum, Nynke L, Meisig, Johannes, Sedat, John, Gribnau, Joost, Barillot, Emmanuel, Blüthgen, Nils, Dekker, Job and Heard, Edith. 2012. Spatial partitioning of the regulatory landscape of the X-inactivation centre. *Nature*, **485**(7398), 381–5.

Nordling, CO. 1953. A new theory on cancer-inducing mechanism. *Br J Cancer*, **7**(1), 68–72.

Nowak, MA. 2006. *Evolutionary Dynamics: Exploring the Equations of Life*. Harvard University Press.

Nowell, PC. 1976. The clonal evolution of tumor cell populations. *Science*, **194**(4260), 23–8.

Nugent, LJ and Jain, RK. 1984. Extravascular diffusion in normal and neoplastic tissues. *Cancer Res*, **44**(1), 238–44.

Oakes, Patrick W, Patel, Dipan C, Morin, Nicole A, Zitterbart, Daniel P, Fabry, Ben, Reichner, Jonathan S and Tang, Jay X. 2009. Neutrophil morphology and migration are affected by substrate elasticity. *Blood*, **114**(7), 1387–95.

O'Connell, Christopher B and Khodjakov, Alexey L. 2007. Cooperative mechanisms of mitotic spindle formation. *J Cell Sci*, **120**(Pt 10), 1717–22.

Oh, SJ, Jeltsch, MM, Birkenhäger, R, McCarthy, JE, Weich, HA, Christ, B, Alitalo, K and Wilting, J. 1997. VEGF and VEGF-C: specific induction of angiogenesis and lymphangiogenesis in the differentiated avian chorioallantoic membrane. *Dev Biol*, **188**(1), 96–109.

Oren, M. 2003. Decision making by p53: life, death and cancer. *Cell Death Differ*, **10**(4), 431–42.

Pacardo, Dennis B, Ligler, Frances S and Gu, Zhen. 2015. Programmable nanomedicine: synergistic and sequential drug delivery systems. *Nanoscale*, **7**(8), 3381–91.

Padera, Timothy P, Stoll, Brian R, Tooredman, Jessica B, Capen, Diane, di Tomaso, Emmanuelle and Jain, Rakesh K. 2004. Pathology: cancer cells compress intratumour vessels. *Nature*, **427**(6976), 695.

Paluch, Ewa K and Raz, Erez. 2013. The role and regulation of blebs in cell migration. *Curr Opin Cell Biol*, **25**(5), 582–90.

Parent, CA and Devreotes, PN. 1999. A cell's sense of direction. *Science*, **284**(5415), 765–70.

Park, Jin-Ah, Kim, Jae Hun, Bi, Dapeng, Mitchel, Jennifer A, Qazvini, Nader Taheri, Tantisira, Kelan, Park, Chan Young, McGill, Maureen, Kim, Sae-Hoon, Gweon,

Bomi, Notbohm, Jacob, Steward, Jr, Robert, Burger, Stephanie, Randell, Scott H, Kho, Alvin T, Tambe, Dhananjay T, Hardin, Corey, Shore, Stephanie A, Israel, Elliot, Weitz, David A, Tschumperlin, Daniel J, Henske, Elizabeth P, Weiss, Scott T, Manning, M Lisa, Butler, James P, Drazen, Jeffrey M and Fredberg, Jeffrey J. 2015. Unjamming and cell shape in the asthmatic airway epithelium. *Nat Mater*, Aug.

Park, So Yeon, Lee, Hee Eun, Li, Hailun, Shipitsin, Michail, Gelman, Rebecca and Polyak, Kornelia. 2010. Heterogeneity for stem cell-related markers according to tumor subtype and histologic stage in breast cancer. *Clin Cancer Res*, **16**(3), 876–87.

Pasquale, Elena B. 2010. Eph receptors and ephrins in cancer: bidirectional signalling and beyond. *Nat Rev Cancer*, **10**(3), 165–80.

Paszek, Matthew J and Weaver, Valerie M. 2004. The tension mounts: mechanics meets morphogenesis and malignancy. *Journal of Mammary Gland Biology and Neoplasia*, **9**(4), 325–42.

Paszek, Matthew J, Zahir, Nastaran, Johnson, Kandice R, Lakins, Johnathon N, Rozenberg, Gabriela I, Gefen, Amit, Reinhart-King, Cynthia A, Margulies, Susan S, Dembo, Micah, Boettiger, David, et al. 2005. Tensional homeostasis and the malignant phenotype. *Cancer Cell*, **8**(3), 241–254.

Paul, Raja, Wollman, Roy, Silkworth, William T., Nardi, Isaac K., Cimini, Daniela and Mogilner, Alex. 2009. Computer simulations predict that chromosome movements and rotations accelerate mitotic spindle assembly without compromising accuracy. *PNAS*, **106**, 15708–13.

Peer, Dan, Karp, Jeffrey M, Hong, Seungpyo, Farokhzad, Omid C, Margalit, Rimona and Langer, Robert. 2007. Nanocarriers as an emerging platform for cancer therapy. *Nat Nanotechnol*, **2**(12), 751–60.

Perelson, Alan S and Weisbuch, Gérard. 1997. Immunology for physicists. *Rev. Mod. Phys.*, **69**(Oct), 1219–68.

Phillips, Rob. 2015. Theory in Biology: Figure 1 or Figure 7? *Trends Cell Biol*, **25**(12), 723–9.

Piano, Amanda, and Titorenko, Vladimir I. 2015. The Intricate Interplay between Mechanisms Underlying Aging and Cancer. *Aging Dis*, **6**(1), 56–75.

Poirier, MG and Marko, JF. 2002. Micromechanical studies of mitotic chromosomes. *J Muscle Res Cell Motil*, **23**(5-6), 409–31.

Potdar, Alka A, Jeon, Junhwan, Weaver, Alissa M, Quaranta, Vito and Cummings, Peter T. 2010. Human mammary epithelial cells exhibit a bimodal correlated random walk pattern. *PLoS One*, **5**(3), e9636.

Potten, CS and Morris, RJ. 1988. Epithelial stem cells in vivo. *J Cell Sci Suppl*, **10**, 45–62.

Poujade, M, Grasland-Mongrain, E, Hertzog, A, Jouanneau, J, Chavrier, P, Ladoux, B, Buguin, A and Silberzan, P. 2007. Collective migration of an epithelial monolayer in response to a model wound. *Proc Natl Acad Sci U S A*, **104**(41), 15988–93.

Prasmickaite, Lina, Engesaeter, Birgit Ø, Skrbo, Nirma, Hellenes, Tina, Kristian, Alexandr, Oliver, Nina K, Suo, Zhenhe and Maelandsmo, Gunhild M. 2010. Aldehyde dehydrogenase (ALDH) activity does not select for cells with enhanced aggressive properties in malignant melanoma. *PLoS One*, **5**(5), e10731.

Prevo, R, Banerji, S, Ferguson, D J, Clasper, S and Jackson, D G. 2001. Mouse LYVE-1 is an endocytic receptor for hyaluronan in lymphatic endothelium. *J Biol Chem*, **276**(22), 19420–30.

Prost, J, Julicher, F and Joanny, J-F. 2015. Active gel physics. *Nat Phys*, **11**(2), 111–17.

Provenzano, Paolo P, Eliceiri, Kevin W, Campbell, Jay M, Inman, David R, White, John G and Keely, Patricia J. 2006. Collagen reorganization at the tumor-stromal interface facilitates local invasion. *BMC Medicine*, **4**(1), 38.

Provenzano, Paolo P, Inman, David R, Eliceiri, Kevin W, Knittel, Justin G, Yan, Long, Rueden, Curtis T, White, John G and Keely, Patricia J. 2008. Collagen density promotes mammary tumor initiation and progression. *BMC Medicine*, **6**(1), 11.

Putkey, Frances R., Cramer, Thorsten, Morphew, Mary K, Silk, Alain D, Johnson, Randall S, McIntosh, J. Richard and Cleveland, Don W. 2002. Unstable Kinetochore-Microtubule Capture and Chromosomal Instability Following Deletion of CENP-E. *Dev. Cell*, **3**, 351–65.

Quintana, Elsa, Shackleton, Mark, Sabel, Michael S, Fullen, Douglas R, Johnson, Timothy M and Morrison, Sean J. 2008. Efficient tumour formation by single human melanoma cells. *Nature*, **456**(7222), 593–8.

Quintana, Elsa, Shackleton, Mark, Foster, Hannah R, Fullen, Douglas R, Sabel, Michael S, Johnson, Timothy M and Morrison, Sean J. 2010. Phenotypic heterogeneity among tumorigenic melanoma cells from patients that is reversible and not hierarchically organized. *Cancer Cell*, **18**(5), 510–23.

Raab, M, Gentili, M, de Belly, H, Thiam, HR, Vargas, P, Jimenez, AJ, Lautenschlaeger, F, Voituriez, Raphaël, Lennon-Duménil, AM, Manel, N and Piel, M. 2016. ESCRT III repairs nuclear envelope ruptures during cell migration to limit DNA damage and cell death. *Science*, **352**(6283), 359–62.

Rakshit, Sabyasachi, Zhang, Yunxiang, Manibog, Kristine, Shafraz, Omer and Sivasankar, Sanjeevi. 2012. Ideal, catch, and slip bonds in cadherin adhesion. *Proc Natl Acad Sci U S A*, **109**(46), 18815–20.

Ramachandran, Vishwanath, Williams, Marcie, Yago, Tadayuki, Schmidtke, David W and McEver, Rodger P. 2004. Dynamic alterations of membrane tethers stabilize leukocyte rolling on P-selectin. *Proc Natl Acad Sci U S A*, **101**(37), 13519–24.

Ramon y Cajal, Santiago. 1999. *Advice for a Young Investigator*. Cambridge, MA: The MIT Press.

Rappel, Wouter-Jan and Levine, Herbert. 2008a. Receptor noise and directional sensing in eukaryotic chemotaxis. *Phys Rev Lett*, **100**(22), 228101.

Rappel, Wouter-Jan and Levine, Herbert. 2008b. Receptor noise limitations on chemotactic sensing. *Proc Natl Acad Sci U S A*, **105**(49), 19270–5.

Raza, Ahmad, Franklin, Michael J and Dudek, Arkadiusz Z. 2010. Pericytes and vessel maturation during tumor angiogenesis and metastasis. *Am J Hematol*, **85**(8), 593–8.

Reddy, Karen L and Feinberg, Andrew P. 2013. Higher order chromatin organization in cancer. *Semin Cancer Biol*, **23**(2), 109–15.

Remmerbach, Torsten W, Wottawah, Falk, Dietrich, Julia, Lincoln, Bryan, Wittekind, Christian and Guck, Jochen. 2009. Oral cancer diagnosis by mechanical phenotyping. *Cancer Research*, **69**(5), 1728–32.

Ricart, Brendon G, Yang, Michael T, Hunter, Christopher A, Chen, Christopher S and Hammer, Daniel A. 2011. Measuring traction forces of motile dendritic cells on micropost arrays. *Biophys J*, **101**(11), 2620–8.

Rich, Alexander and Zhang, Shuguang. 2003. Timeline: Z-DNA: the long road to biological function. *Nat Rev Genet*, **4**(7), 566–72.

Rieder, CL and Alexander, SP. 1990. Kinetochores are transported poleward along a single astral microtubule during chromosome attachment to the spindle in newt lung cells. *J. Cell Biol.*, **110**, 81–95.

Rieder, CL, Davison, EA, Jensen, LC, Cassimeris, L and Salmon, ED. 1986. Oscillatory movements of monooriented chromosomes and their position relative to the spindle pole result from the ejection properties of the aster and half-spindle. *J Cell Biol*, **103**(2), 581–91.

Rieger, Heiko and Welter, Michael. 2015. Integrative models of vascular remodeling during tumor growth. *Wiley Interdisciplinary Reviews: Systems Biology and Medicine*, **7**(3), 113–29.

Risca, Viviana I and Greenleaf, William J. 2015. Unraveling the 3D genome: genomics tools for multiscale exploration. *Trends Genet*, **31**(7), 357–72.

Robertson-Tessi, Mark, El-Kareh, Ardith and Goriely, Alain. 2012. A mathematical model of tumor-immune interactions. *J Theor Biol*, **294**(Feb), 56–73.

Rodier, Francis and Campisi, Judith. 2011. Four faces of cellular senescence. *J Cell Biol*, **192**(4), 547–56.

Roesch, Alexander, Fukunaga-Kalabis, Mizuho, Schmidt, Elizabeth C, Zabierowski, Susan E, Brafford, Patricia A, Vultur, Adina, Basu, Devraj, Gimotty, Phyllis, Vogt, Thomas and Herlyn, Meenhard. 2010. A temporarily distinct subpopulation of slow-cycling melanoma cells is required for continuous tumor growth. *Cell*, **141**(4), 583–94.

Rofstad, Einar K, Tunheim, Siv H, Mathiesen, Berit, Graff, Bjørn A, Halsør, Ellen F, Nilsen, Kristin and Galappathi, Kanthi. 2002. Pulmonary and lymph node metastasis is associated with primary tumor interstitial fluid pressure in human melanoma xenografts. *Cancer Research*, **62**(3), 661–64.

Rørth, Pernille. 2009. Collective cell migration. *Annu Rev Cell Dev Biol*, **25**, 407–29.

Rosoff, William J, Urbach, Jeffrey S, Esrick, Mark A, McAllister, Ryan G, Richards, Linda J and Goodhill, Geoffrey J. 2004. A new chemotaxis assay shows the extreme sensitivity of axons to molecular gradients. *Nat Neurosci*, **7**(6), 678–82.

Ross, Sharon A and Davis, Cindy D. 2011. MicroRNA, nutrition, and cancer prevention. *Adv Nutr*, **2**(6), 472–85.

Röttgermann, Peter JF, Alberola, Alicia Piera and Rädler, Joachim O. 2014. Cellular self-organization on micro-structured surfaces. *Soft Matter*, **10**(14), 2397–404.

Rouzina, I and Bloomfield, VA. 1998. DNA bending by small, mobile multivalent cations. *Biophys J*, **74**(6), 3152–64.

Ruf, Wolfram, Seftor, Elisabeth A, Petrovan, Ramona J, Weiss, Robert M, Gruman, Lynn M, Margaryan, Naira V, Seftor, Richard EB, Miyagi, Yohei and Hendrix, Mary JC. 2003. Differential role of tissue factor pathway inhibitors 1 and 2 in melanoma vasculogenic mimicry. *Cancer Res*, **63**(17), 5381–9.

Ruoslahti, Erkki, Bhatia, Sangeeta N and Sailor, Michael J. 2010. Targeting of drugs and nanoparticles to tumors. *J Cell Biol*, **188**(6), 759–68.

Ruprecht, Verena, Wieser, Stefan, Callan-Jones, Andrew, Smutny, Michael, Morita, Hitoshi, Sako, Keisuke, Barone, Vanessa, Ritsch-Marte, Monika, Sixt, Michael, Voi-turiez, Raphaël and Heisenberg, Carl-Philipp. 2015. Cortical contractility triggers a stochastic switch to fast amoeboid cell motility. *Cell*, **160**(4), 673–85.

Salbreux, G, Prost, J and Joanny, JF. 2009. Hydrodynamics of cellular cortical flows and the formation of contractile rings. *Phys Rev Lett*, **103**(5), 058102.

Salmaggi, Andrea, Boiardi, Amerigo, Gelati, Maurizio, Russo, Annamaria, Calatozzolo, Chiara, Ciusani, Emilio, Sciacca, Francesca Luisa, Ottolina, Arianna, Parati, Euge-nio Agostino, La Porta, Caterina, Alessandri, Giulio, Marras, Carlo, Croci, Danilo and De Rossi, Marco. 2006. Glioblastoma-derived tumorospheres identify a population of tumor stem-like cells with angiogenic potential and enhanced multidrug resistance phenotype. *Glia*, **54**(8), 850–60.

Salnikov, Alexei V, Iversen, Vegard V, Koisti, Markus, Sundberg, Christian, Johansson, Lars, Stuhr, Linda B, Sjöquist, Mats, Ahlström, Håkan, Reed, Rolf K and Rubin,

Kristofer. 2003. Lowering of tumor interstitial fluid pressure specifically augments efficacy of chemotherapy. *The FASEB Journal*, **17**(12), 1756–8.

Saltzman, W Mark and Olbricht, William L. 2002. Building drug delivery into tissue engineering. *Nat Rev Drug Discov*, **1**(3), 177–86.

Samani, Abbas, Bishop, Jonathan, Luginbuhl, Chris and Plewes, Donald B. 2003. Measuring the elastic modulus of ex vivo small tissue samples. *Physics in Medicine and Biology*, **48**(14), 2183.

Sandhu, Sumit, Wu, Xiaoli, Nabi, Zinnatun, Rastegar, Mojgan, Kung, Sam, Mai, Sabine and Ding, Hao. 2012. Loss of HLTF function promotes intestinal carcinogenesis. *Mol Cancer*, **11**, 18.

Sanger, Joseph W. 1975. Changing patterns of actin localization during cell division. *Proceedings of the National Academy of Sciences*, **72**(5), 1913–16.

Sarikaya, Mehmet, Tamerler, Candan, Jen, Alex K-Y, Schulten, Klaus and Baneyx, François. 2003. Molecular biomimetics: nanotechnology through biology. *Nat Mater*, **2**(9), 577–85.

Sasikala, Arathyram Ramachandra Kurup, GhavamiNejad, Amin, Unnithan, Afeesh Rajan, Thomas, Reju George, Moon, Myeongju, Jeong, Yong Yeon, Park, Chan Hee and Kim, Cheol Sang. 2015. A smart magnetic nanoplatform for synergistic anticancer therapy: manoeuvring mussel-inspired functional magnetic nanoparticles for pH responsive anticancer drug delivery and hyperthermia. *Nanoscale*, **7**(43), 18119–28.

Schatton, Tobias, Murphy, George F, Frank, Natasha Y, Yamaura, Kazuhiro, Waaga-Gasser, Ana Maria, Gasser, Martin, Zhan, Qian, Jordan, Stefan, Duncan, Lyn M, Weishaupt, Carsten, Fuhlbrigge, Robert C, Kupper, Thomas S, Sayegh, Mohamed H and Frank, Markus H. 2008. Identification of cells initiating human melanomas. *Nature*, **451**(7176), 345–9.

Schepers, Arnout G, Snippert, Hugo J, Stange, Daniel E, van den Born, Maaike, van Es, Johan H, van de Wetering, Marc and Clevers, Hans. 2012. Lineage tracing reveals Lgr5+ stem cell activity in mouse intestinal adenomas. *Science*, **337**(6095), 730–5.

Schmid, Michael C and Varner, Judith A. 2010. Myeloid cells in the tumor microenvironment: modulation of tumor angiogenesis and tumor inflammation. *J Oncol*, **2010**, 201026.

Schmidt, Thorsten I, Kleylein-Sohn, Julia, Westendorf, Jens, Le Clech, Mikael, Lavoie, Sébastien B, Stierhof, York-Dieter and Nigg, Erich A. 2009. Control of centriole length by CPAP and CP110. *Curr Biol*, **19**(12), 1005–11.

Schoppmann, Sebastian F, Birner, Peter, Stöckl, Johannes, Kalt, Romana, Ullrich, Robert, Caucig, Carola, Kriehuber, Ernst, Nagy, Katalin, Alitalo, Kari and Kerjaschki, Dontscho. 2002. Tumor-associated macrophages express lymphatic endothelial growth factors and are related to peritumoral lymphangiogenesis. *Am J Pathol*, **161**(3), 947–56.

Sellerio, Alessandro L, Ciusani, Emilio, Ben-Moshe, Noa Bossel, Coco, Stefania, Piccinini, Andrea, Myers, Christopher R, Sethna, James P, Giampietro, Costanza, Zapperi, Stefano and La Porta, Caterina A M. 2015. Overshoot during phenotypic switching of cancer cell populations. *Sci Rep*, **5**, 15464.

Sepúlveda, Néstor, Petitjean, Laurence, Cochet, Olivier, Grasland-Mongrain, Erwan, Silberzan, Pascal and Hakim, Vincent. 2013. Collective Cell Motion in an Epithelial Sheet Can Be Quantitatively Described by a Stochastic Interacting Particle Model. *PLoS Comput Biol*, **9**(3), e1002944.

Serra, R, Villani, M and Semeria, A. 2004. Genetic network models and statistical properties of gene expression data in knock-out experiments. *J Theor Biol*, **227**(1), 149–57.

Serra, R, Villani, M, Graudenzi, A, and Kauffman, SA. 2007. Why a simple model of genetic regulatory networks describes the distribution of avalanches in gene expression data. *J Theor Biol*, **246**(3), 449–60.

Serra-Picamal, Xavier, Conte, Vito, Vincent, Romaric, Anon, Ester, Tambe, Dhananjay T, Bazellieres, Elsa, Butler, James P, Fredberg, Jeffrey J and Trepat, Xavier. 2012. Mechanical waves during tissue expansion. *Nat Phys*, **8**(8), 628–34.

Shanahan, Joseph C and St Clair, William. 2002. Tumor necrosis factor-alpha blockade: a novel therapy for rheumatic disease. *Clin Immunol*, **103**(3 Pt 1), 231–42.

Sharp, Phillip A and Langer, Robert. 2011. Research agenda. Promoting convergence in biomedical science. *Science*, **333**(6042), 527.

Shen, Michael M and Abate-Shen, Cory. 2010. Molecular genetics of prostate cancer: new prospects for old challenges. *Genes Dev*, **24**(18), 1967–2000.

Shi, Jinjun, Votruba, Alexander R, Farokhzad, Omid C and Langer, Robert. 2010. Nanotechnology in drug delivery and tissue engineering: from discovery to applications. *Nano Lett*, **10**(9), 3223–30.

Shimo, Arata, Tanikawa, Chizu, Nishidate, Toshihiko, Lin, Meng-Lay, Matsuda, Koichi, Park, Jae-Hyun, Ueki, Tomomi, Ohta, Tomohiko, Hirata, Koichi, Fukuda, Mamoru, Nakamura, Yusuke and Katagiri, Toyomasa. 2008. Involvement of kinesin family member 2C/mitotic centromere-associated kinesin overexpression in mammary carcinogenesis. *Cancer Sci*, **99**(1), 62–70.

Shklovskii, BI. 1999. Screening of a macroion by multivalent ions: correlation-induced inversion of charge. *Phys Rev E Stat Phys Plasmas Fluids Relat Interdiscip Topics*, **60**(5 Pt B), 5802–11.

Silk, Alain D, Zasadil, Lauren M, Holland, Andrew J, Vitre, Benjamin, Cleveland, Don W and Weaver, Beth A. 2013. Chromosome missegregation rate predicts whether aneuploidy will promote or suppress tumors. *Proc Natl Acad Sci U S A*, **110**(44), E4134–41.

Skalak, Richard, Zargaryan, Stephen, Jain, Rakesh K, Netti, Paolo A and Hoger, Anne. 1996. Compatibility and the genesis of residual stress by volumetric growth. *Journal of Mathematical Biology*, **34**(8), 889–914.

Sleeman, JP, Krishnan, J, Kirkin, V and Baumann, P. 2001. Markers for the lymphatic endothelium: in search of the holy grail? *Microsc Res Tech*, **55**(2), 61–9.

Smith, Douglas M, Simon, Jakub K and Baker, Jr, James R. 2013. Applications of nanotechnology for immunology. *Nat Rev Immunol*, **13**(8), 592–605.

Söllner, T, Whiteheart, SW, Brunner, M, Erdjument-Bromage, H, Geromanos, S, Tempst, P and Rothman, JE. 1993. SNAP receptors implicated in vesicle targeting and fusion. *Nature*, **362**(6418), 318–24.

Sosa-Pineda, B, Wigle, JT and Oliver, G. 2000. Hepatocyte migration during liver development requires Prox1. *Nat Genet*, **25**(3), 254–5.

Sottoriva, Andrea, Verhoeff, Joost JC, Borovski, Tijana, McWeeney, Shannon K, Naumov, Lev, Medema, Jan Paul, Sloot, Peter MA and Vermeulen, Louis. 2010. Cancer stem cell tumor model reveals invasive morphology and increased phenotypical heterogeneity. *Cancer Res*, **70**(1), 46–56.

Sprouffske, Kathleen, Pepper, John W and Maley, Carlo C. 2011. Accurate reconstruction of the temporal order of mutations in neoplastic progression. *Cancer Prev Res (Phila)*, **4**(7), 1135–44.

Steinway, Steven Nathaniel, Zañudo, Jorge GT, Ding, Wei, Rountree, Carl Bart, Feith, David J, Loughran, Jr, Thomas P and Albert, Reka. 2014. Network modeling of TGF signaling in hepatocellular carcinoma epithelial-to-mesenchymal transition reveals joint sonic hedgehog and Wnt pathway activation. *Cancer Res*, **74**(21), 5963–77.

Stokes, CL, Lauffenburger, DA and Williams, SK. 1991. Migration of individual microvessel endothelial cells: stochastic model and parameter measurement. *J Cell Sci*, **99 (Pt 2)**(Jun), 419–30.

Stout, Jane R, Rizk, Rania S, Kline, Susan L and Walczak, Claire E. 2006. Deciphering protein function during mitosis in PtK cells using RNAi. *BMC Cell Biol*, **7**, 26.

Stroka, Kimberly M, Jiang, Hongyuan, Chen, Shih-Hsun, Tong, Ziqiu, Wirtz, Denis, Sun, Sean X and Konstantopoulos, Konstantinos. 2014. Water permeation drives tumor cell migration in confined microenvironments. *Cell*, **157**(3), 611–23.

Strukov, Yuri G and Belmont, AS. 2009. Mitotic chromosome structure: reproducibility of folding and symmetry between sister chromatids. *Biophys J*, **96**(4), 1617–28.

Stumpff, Jason, Wagenbach, Michael, Franck, Andrew, Asbury, Charles L and Wordeman, Linda. 2012. Kif18A and chromokinesins confine centromere movements via microtubule growth suppression and spatial control of kinetochore tension. *Dev Cell*, **22**(5), 1017–29.

Stylianopoulos, Triantafyllos. 2013. EPR-effect: utilizing size-dependent nanoparticle delivery to solid tumors. *Ther Deliv*, **4**(4), 421–3.

Stylianopoulos, Triantafyllos, Martin, John D, Chauhan, Vikash P, Jain, Saloni R, Diop-Frimpong, Benjamin, Bardeesy, Nabeel, Smith, Barbara L, Ferrone, Cristina R, Hornicek, Francis J, Boucher, Yves, et al. 2012. Causes, consequences, and remedies for growth-induced solid stress in murine and human tumors. *Proceedings of the National Academy of Sciences*, **109**(38), 15101–15108.

Stylianopoulos, Triantafyllos, Martin, John D, Snuderl, Matija, Mpekris, Fotios, Jain, Saloni R and Jain, Rakesh K. 2013. Coevolution of solid stress and interstitial fluid pressure in tumors during progression: Implications for vascular collapse. *Cancer research*, **73**(13), 3833–41.

Suresh, Subra. 2007. Biomechanics and biophysics of cancer cells. *Acta Materialia*, **55**(12), 3989–4014.

Szabó, B, Szöllösi, GJ, Gönci, B, Jurányi, Zs, Selmeczi, D and Vicsek, Tamás. 2006. Phase transition in the collective migration of tissue cells: Experiment and model. *Phys. Rev. E*, **74**(Dec), 061908.

Taghizadeh, Rouzbeh, Noh, Minsoo, Huh, Yang Hoon, Ciusani, Emilio, Sigalotti, Luca, Maio, Michele, Arosio, Beatrice, Nicotra, Maria R, Natali, PierGiorgio, Sherley, James L and La Porta, Caterina AM. 2010. CXCR6, a newly defined biomarker of tissue-specific stem cell asymmetric self-renewal, identifies more aggressive human melanoma cancer stem cells. *PLoS One*, **5**(12), e15183.

Taloni, Alessandro, Alemi, Alexander A, Ciusani, Emilio, Sethna, James P, Zapperi, Stefano and La Porta, Caterina AM. 2014. Mechanical properties of growing melanocytic nevi and the progression to melanoma. *PLoS One*, **9**(4), e94229.

Taloni, Alessandro, Ben Amar, Martine, Zapperi, Stefano and La Porta, Caterina AM. 2015a. The role of pressure in cancer growth. *The European Physical Journal Plus*, **130**(11).

Taloni, Alessandro, Kardash, Elena, Salman, Oguz Umut, Truskinovsky, Lev, Zapperi, Stefano and La Porta, Caterina AM. 2015b. Volume Changes During Active Shape Fluctuations in Cells. *Phys Rev Lett*, **114**(20), 208101.

Tambe, Dhananjay T, Hardin, C Corey, Angelini, Thomas E, Rajendran, Kavitha, Park, Chan Young, Serra-Picamal, Xavier, Zhou, Enhua H, Zaman, Muhammad H, Butler, James P, Weitz, David A, Fredberg, Jeffrey J and Trepat, Xavier. 2011. Collective cell guidance by cooperative intercellular forces. *Nat Mater*, **10**(6), 469–75.

Tammela, Tuomas and Alitalo, Kari. 2010. Lymphangiogenesis: Molecular mechanisms and future promise. *Cell*, **140**(4), 460–76.

Tanenbaum, Marvin E, Macurek, Libor, van der Vaart, Babet, Galli, Matilde, Akhmanova, Anna and Medema, René H. 2011. A complex of Kif18b and MCAK promotes microtubule depolymerization and is negatively regulated by Aurora kinases. *Curr Biol*, **21**(16), 1356–65.

Tang, Dean G, Patrawala, Lubna, Calhoun, Tammy, Bhatia, Bobby, Choy, Grace, Schneider-Broussard, Robin and Jeter, Collene. 2007. Prostate cancer stem/progenitor cells: identification, characterization, and implications. *Mol Carcinog*, **46**(1), 1–14.

Tao, Yu, Ju, Enguo, Liu, Zhen, Dong, Kai, Ren, Jinsong and Qu, Xiaogang. 2014. Engineered, self-assembled near-infrared photothermal agents for combined tumor immunotherapy and chemo-photothermal therapy. *Biomaterials*, **35**(24), 6646–56.

Théry, Manuel. 2010. Micropatterning as a tool to decipher cell morphogenesis and functions. *J Cell Sci*, **123**(Pt 24), 4201–13.

Thomas, Wendy. 2008. Catch bonds in adhesion. *Annu Rev Biomed Eng*, **10**, 39–57.

Thomas, Wendy E, Vogel, Viola and Sokurenko, Evgeni. 2008. Biophysics of catch bonds. *Annu Rev Biophys*, **37**, 399–416.

Thompson, Sarah L, Bakhoum, Samuel F and Compton, Duane A. 2010. Mechanisms of chromosomal instability. *Curr Biol*, **20**(6), R285–95.

Tiana, G, Krishna, S, Pigolotti, S, Jensen, MH and Sneppen, K. 2007. Oscillations and temporal signalling in cells. *Phys Biol*, **4**(2), R1–17.

Tiana, Guido, Jensen, MH and Sneppen, Kim. 2002. Time delay as a key to apoptosis induction in the p53 network. *The European Physical Journal B-Condensed Matter and Complex Systems*, **29**(1), 135–40.

Tolhuis, Bas, Palstra, Robert Jan, Splinter, Erik, Grosveld, Frank and de Laat, Wouter. 2002. Looping and interaction between hypersensitive sites in the active beta-globin locus. *Mol Cell*, **10**(6), 1453–65.

Tomasetti, Cristian and Levy, Doron. 2010. Role of symmetric and asymmetric division of stem cells in developing drug resistance. *Proc Natl Acad Sci U S A*, **107**(39), 16766–71.

Tomasetti, Cristian and Vogelstein, Bert. 2015. Cancer etiology. Variation in cancer risk among tissues can be explained by the number of stem cell divisions. *Science*, **347**(6217), 78–81.

Tower, John. 2015. Programmed cell death in aging. *Ageing Res Rev*, **23**(Pt A), 90–100.

Trepat, Xavier, Wasserman, Michael R, Angelini, Thomas E, Millet, Emil, Weitz, David A, Butler, James P and Fredberg, Jeffrey J. 2009. Physical forces during collective cell migration. *Nat Phys*, **5**(6), 426–430.

Turchinovich, Andrey, Weiz, Ludmila, Langheinz, Anne and Burwinkel, Barbara. 2011. Characterization of extracellular circulating microRNA. *Nucleic Acids Res*, **39**(16), 7223–33.

Valadi, Hadi, Ekström, Karin, Bossios, Apostolos, Sjöstrand, Margareta, Lee, James J and Lötvall, Jan O. 2007. Exosome-mediated transfer of mRNAs and microRNAs is a novel mechanism of genetic exchange between cells. *Nat Cell Biol*, **9**(6), 654–9.

Van Haastert, Peter JM and Veltman, Douwe M. 2007. Chemotaxis: navigating by multiple signaling pathways. *Sci STKE*, **2007**(396), pe40.

Vanharanta, Sakari, Shu, Weiping, Brenet, Fabienne, Hakimi, A Ari, Heguy, Adriana, Viale, Agnes, Reuter, Victor E, Hsieh, James J-D, Scandura, Joseph M and Massagué, Joan. 2013. Epigenetic expansion of VHL-HIF signal output drives multiorgan metastasis in renal cancer. *Nat Med*, **19**(1), 50–6.

Vedula, Sri Ram Krishna, Ravasio, Andrea, Lim, Chwee Teck and Ladoux, Benoit. 2013. Collective cell migration: a mechanistic perspective. *Physiology (Bethesda)*, **28**(6), 370–9.

Vial, Emmanuel, Sahai, Erik and Marshall, Christopher J. 2003. ERK-MAPK signaling coordinately regulates activity of Rac1 and RhoA for tumor cell motility. *Cancer Cell*, **4**(1), 67–79.

Vogel, H, Niewisch, H and Matioli, G. 1968. The self renewal probability of hemopoietic stem cells. *J Cell Physiol*, **72**(3), 221–8.

Vogelstein, B, Lane, D and Levine, AJ. 2000. Surfing the p53 network. *Nature*, **408**(6810), 307–10.

Voituriez, R, Joanny, JF and Prost, J. 2006. Generic phase diagram of active polar films. *Phys Rev Lett*, **96**(2), 028102.

von Maltzahn, Geoffrey, Park, Ji-Ho, Agrawal, Amit, Bandaru, Nanda Kishor, Das, Sarit K, Sailor, Michael J and Bhatia, Sangeeta N. 2009. Computationally guided photothermal tumor therapy using long-circulating gold nanorod antennas. *Cancer Res*, **69**(9), 3892–900.

Vorozhko, VV, Emanuele, MJ, Kallio, MJ and Gorbsky, PT, Stukenberg GJ. 2008. Multiple mechanisms of chromosome movement in vertebrate cells mediated through the Ndc80 complex and dynein/dynactin. *Chromosoma*, **117**, 169–79.

Waclaw, Bartlomiej, Bozic, Ivana, Pittman, Meredith E, Hruban, Ralph H, Vogelstein, Bert and Nowak, Martin A. 2015. A spatial model predicts that dispersal and cell turnover limit intratumour heterogeneity. *Nature*, Aug.

Wagner, Volker, Dullaart, Anwyn, Bock, Anne-Katrin and Zweck, Axel. 2006. The emerging nanomedicine landscape. *Nat Biotechnol*, **24**(10), 1211–7.

Walczak, Claire E, Cai, Shang and Khodjakov, Alexey. 2010. Mechanisms of chromosome behaviour during mitosis. *Nat Rev Mol Cell Biol*, **11**(2), 91–102.

Walker, DC, Southgate, J, Hill, G, Holcombe, M, Hose, DR, Wood, SM, MacNeil, S and Smallwood, RH. 2004. The epitheliome: agent-based modelling of the social behaviour of cells. *Biosystems*, **76**(1-3), 89–100.

Wandke, C, Barisic, M, Sigl, R, Rauch, V, Wolf, F, Amaro, AC, Tan, CH, Pereira, AJ, Kutay, U, Maiato, H, Meraldi, P and Geley, S. 2012. Human chromokinesins promote chromosome congression and spindle microtubule dynamics during mitosis. *J. Cell Biol.*, **198**, 847–63.

Wang, Xi, Kruithof-de Julio, Marianna, Economides, Kyriakos D, Walker, David, Yu, Hailong, Halili, M Vivienne, Hu, Ya-Ping, Price, Sandy M, Abate-Shen, Cory and Shen, Michael M. 2009. A luminal epithelial stem cell that is a cell of origin for prostate cancer. *Nature*, **461**(7263), 495–500.

Watson, JD and Crick, FH. 1953. Molecular structure of nucleic acids; a structure for deoxyribose nucleic acid. *Nature*, **171**(4356), 737–8.

Weiner, Orion D. 2002. Regulation of cell polarity during eukaryotic chemotaxis: the chemotactic compass. *Curr Opin Cell Biol*, **14**(2), 196–202.

Wells, KE, Rapaport, DP, Cruse, CW, Payne, W, Albertini, J, Berman, C, Lyman, GH and Reintgen, DS. 1997. Sentinel lymph node biopsy in melanoma of the head and neck. *Plast Reconstr Surg*, **100**(3), 591–4.

Wells, Rebecca G. 2005. The role of matrix stiffness in hepatic stellate cell activation and liver fibrosis. *Journal of Clinical Gastroenterology*, **39**(4), S158–S161.

Welter, M and Rieger, H. 2010. Physical determinants of vascular network remodeling during tumor growth. *The European Physical Journal E*, **33**(2), 149–63.

Welter, Michael and Rieger, Heiko. 2013. Interstitial fluid flow and drug delivery in vascularized tumors: a computational model. *PloS One*, **8**(8), e70395.

Weninger, W, Partanen, TA, Breiteneder-Geleff, S, Mayer, C, Kowalski, H, Mildner, M, Pammer, J, Stürzl, M, Kerjaschki, D, Alitalo, K and Tschachler, E. 1999. Expression of vascular endothelial growth factor receptor-3 and podoplanin suggests a lymphatic endothelial cell origin of Kaposi's sarcoma tumor cells. *Lab Invest*, **79**(2), 243–51.

Whipple, Rebecca A, Matrone, Michael A, Cho, Edward H, Balzer, Eric M, Vitolo, Michele I, Yoon, Jennifer R, Ioffe, Olga B, Tuttle, Kimberly C, Yang, Jing and Martin, Stuart S. 2010. Epithelial-to-mesenchymal transition promotes tubulin detyrosination and microtentacles that enhance endothelial engagement. *Cancer Res*, **70**(20), 8127–37.

White, Eileen and DiPaola, Robert S. 2009. The double-edged sword of autophagy modulation in cancer. *Clin Cancer Res*, **15**(17), 5308–16.

Whitehead, Joanne, Vignjevic, Danijela, Fütterer, Claus, Beaurepaire, Emmanuel, Robine, Sylvie and Farge, Emmanuel. 2008. Mechanical factors activate ß-catenin-dependent oncogene expression in APC1638N/+ mouse colon. *HFSP Journal*, **2**(5), 286–94.

Wigle, JT and Oliver, G. 1999. Prox1 function is required for the development of the murine lymphatic system. *Cell*, **98**(6), 769–78.

Wigle, JT, Chowdhury, K, Gruss, P and Oliver, G. 1999. Prox1 function is crucial for mouse lens-fibre elongation. *Nat Genet*, **21**(3), 318–22.

Wirtz, Denis, Konstantopoulos, Konstantinos and Searson, Peter C. 2011. The physics of cancer: the role of physical interactions and mechanical forces in metastasis. *Nat Rev Cancer*, **11**(7), 512–22.

Wodarz, A and Nusse, R. 1998. Mechanisms of Wnt signaling in development. *Annu Rev Cell Dev Biol*, **14**, 59–88.

Wollman, R, Cytrynbaum, EN, Jones, JT, Meyer, T, Scholey, JM and Mogilne, A. 2005. Efficient chromosome capture requires a bias in the 'search-and-capture' process during mitotic-spindle assembly. *Curr. Biol.*, **15**, 828–32.

Wu, Min, Frieboes, Hermann B, McDougall, Steven R, Chaplain, Mark AJ, Cristini, Vittorio and Lowengrub, John. 2013. The effect of interstitial pressure on tumor growth: coupling with the blood and lymphatic vascular systems. *Journal of Theoretical Biology*, **320**, 131–51.

Wu, Pei-Hsun, Giri, Anjil, Sun, Sean X and Wirtz, Denis. 2014. Three-dimensional cell migration does not follow a random walk. *Proc Natl Acad Sci U S A*, **111**(11), 3949–54.

Wu, Song, Powers, Scott, Zhu, Wei and Hannun, Yusuf A. 2015. Substantial contribution of extrinsic risk factors to cancer development. *Nature*, Dec.

Xu, Xiaoyang, Ho, William, Zhang, Xucqing, Bertrand, Nicolas and Farokhzad, Omid. 2015. Cancer nanomedicine: from targeted delivery to combination therapy. *Trends Mol Med*, **21**(4), 223–32.

Yachida, Shinichi, Jones, Siân, Bozic, Ivana, Antal, Tibor, Leary, Rebecca, Fu, Baojin, Kamiyama, Mihoko, Hruban, Ralph H, Eshleman, James R, Nowak, Martin A, Velculescu, Victor E, Kinzler, Kenneth W, Vogelstein, Bert and Iacobuzio-Donahue, Christine A. 2010. Distant metastasis occurs late during the genetic evolution of pancreatic cancer. *Nature*, **467**(7319), 1114–7.

Yang, A, Kaghad, M, Wang, Y, Gillett, E, Fleming, MD, Dötsch, V, Andrews, N C, Caput, D and McKeon, F. 1998. p63, a p53 homolog at 3q27-29, encodes multiple products with transactivating, death-inducing, and dominant-negative activities. *Mol Cell*, **2**(3), 305–16.

Yang, Annie, Kaghad, Mourad, Caput, Daniel and McKeon, Frank. 2002. On the shoulders of giants: p63, p73 and the rise of p53. *Trends Genet*, **18**(2), 90–5.

Yang, Zhenye, Tulu, U Serdar, Wadsworth, Patricia and Rieder, Conly L. 2007. Kinetochore dynein is required for chromosome motion and congression independent of the spindle checkpoint. *Curr Biol*, **17**(11), 973–80.

Zapperi, Stefano and La Porta, Caterina AM. 2012. Do cancer cells undergo phenotypic switching? The case for imperfect cancer stem cell markers. *Sci Rep*, **2**, 441.

Zhang, Nenggang, Ge, Gouquing, Meyer, Rene, Sethi, Sumita, Basu, Dipanjan, Pradhan, Subhashree, Zhao, Yi-Jue, Li, Xiao-Nan, Cai, Wei-Wen, El-Naggar, Adel K, Baladandayuthapani, Veerabhadran, Kittrell, Frances S, Rao, Pulivarthi H, Medina, Daniel and Pati, Debananda. 2008a. Overexpression of Separase induces aneuploidy and mammary tumorigenesis. *Proc Natl Acad Sci U S A*, **105**(35), 13033–8.

Zhang, Ranran, Shah, Mithun Vinod, Yang, Jun, Nyland, Susan B, Liu, Xin, Yun, Jong K, Albert, Réka and Loughran, Jr, Thomas P. 2008b. Network model of survival signaling in large granular lymphocyte leukemia. *Proc Natl Acad Sci U S A*, **105**(42), 16308–13.

Zhang, Xiang H-F, Jin, Xin, Malladi, Srinivas, Zou, Yilong, Wen, Yong H, Brogi, Edi, Smid, Marcel, Foekens, John A and Massagué, Joan. 2013. Selection of bone metastasis seeds by mesenchymal signals in the primary tumor stroma. *Cell*, **154**(5), 1060–73.

Zhu, Changjun, Zhao, Jian, Bibikova, Marina, Leverson, Joel D, Bossy-Wetzel, Ella, Fan, Jian-Bing, Abraham, Robert T and Jiang, Wei. 2005. Functional analysis of human microtubule-based motor proteins, the kinesins and dyneins, in mitosis/cytokinesis using RNA interference. *Mol Biol Cell*, **16**(7), 3187–99.

Zhu, Hongyan, Dougherty, Urszula, Robinson, Victoria, Mustafi, Reba, Pekow, Joel, Kupfer, Sonia, Li, Yan-Chun, Hart, John, Goss, Kathleen, Fichera, Alessandro, Joseph, Loren and Bissonnette, Marc. 2011. EGFR signals downregulate tumor suppressors miR-143 and miR-145 in Western diet-promoted murine colon cancer: role of G1 regulators. *Mol Cancer Res*, **9**(7), 960–75.

Zink, Daniele, Fischer, Andrew H and Nickerson, Jeffrey A. 2004. Nuclear structure in cancer cells. *Nat Rev Cancer*, **4**(9), 677–87.

Index